Explicitly Correlated Wavefunctions

Edited by

Edward A. G. Armour

Jan Franz

Jonathan Tennyson

Explicitly Correlated Wavefunctions

Edited by

Edward A. G. Armour

Department of Mathematics,
University of Nottingham,
University Park,
Nottingham, NG7 2RD,
United Kingdom

Jan Franz

and

Jonathan Tennyson

Department of Physics and Astronomy,
University College London,
Gower Street, London, WC1E6BT,
United Kingdom

Suggested Dewey classification: 539.7, 541.2

ISBN 0-9545289-4-8

Published by

Collaborative Computational Project
on Molecular Quantum Dynamics (CCP6),
Daresbury Laboratory,
Daresbury,
Warrington, WA4 4AD,
United Kingdom

on behalf of

Collaborative Computational Project
on Continuum States of Atoms and Molecules (CCP2),
Daresbury Laboratory,
Daresbury,
Warrington, WA4 4AD,
United Kingdom

Introduction

Collaborative Computing Project 2 (CCP2), supported by EPSRC, is the CCP that has as its area of interest: 'Continuum states of atoms and molecules'. Clearly, this includes multiphoton processes involving atoms and molecules and electron and positron scattering by atoms and molecules. CCP2 is also involved in work on electron scattering by solids, the theory of Bose–Einstein condensation and the interaction of antihydrogen with simple atoms and molecules.

In 2005, Jonathan Tennyson and I were successful in an application to EPSRC for support for a CCP2 flagship project to try to give a definitive explanation for the very high positron annihilation rates that have been observed in positron scattering by molecules such as acteylene (HCCH). See the New Scientist, 24-30 April 2004, page 34. These very high annihilation rates are thought to be to be brought about by a resonant process involving the vibrational motion of the nuclei of the molecule.

The project will combine expertise in CCP2 on accurate calculations on positron-hydrogen-molecule scattering using the Kohn variational method and the application of the R-matrix method to electron scattering by molecules to develop an R-matrix method for carrying out accurate calculations of positron-acetylene scattering. In the past, the R-matrix method has been applied to positron-hydrogen-molecule scattering but this did not give accurate results as the basis sets used were made up of configuration interaction functions that were not able to describe positron-electron correlation adequately.

We have to be able to take this correlation into account accurately if our positron-acetylene calculation is to have any hope of describing the very high annihilation rate process satisfactorily. There is a large amount of expertise amongst quantum chemists on how to take electron-electron correlation into account accurately, in particular by including the interelectron distance as a linear or other suitable factor in the basis sets used in variational calculations.

The aim of the Workshop, of which this booklet is the Proceedings, was to bring together positron physicists and quantum chemists with the expertise just described in treating electron-electron correlation to help us devise an accurate method of taking into account positron-electron correlation using the R-matrix method that can be applied to positron-acetylene scattering, in which the target has fourteen electrons. We are very well pleased with the outcome.

Edward Armour
(Chairman, CCP2)

E. A. G. Armour, J. Franz and J. Tennyson (eds.)
Explicitly Correlated Wavefunctions

Correlations in positron-atom scattering

J.W. Humberston and P. Van Reeth

Department of Physics and Astronomy, University College London, London WC1E 6BT, UK

I. INTRODUCTION

Although the positron, the antiparticle of the electron, has the same mass and intrinsic spin as the electron, its positive charge results in positron-atom interactions being profoundly different from electron-atom interactions. The projectile is now distinguishable from the atomic electrons, so exchange effects are absent. Also, the positron's attraction to the electrons and repulsion by the nucleus makes the direct static interaction with the target atom repulsive whereas the polarization interaction, being quadratic in the charge of the projectile, remains attractive. Thus, two major components of the projectile-atom interaction that are both attractive for electrons are of opposite sign for positrons, resulting in the overall projectile-atom interaction being less attractive for positrons than for electrons. The total scattering cross section is therefore usually smaller for positrons than for electrons, particularly at low projectile energies.

It is a popular misconception that the absence of exchange makes the calculation of scattering parameters easier for positrons than for electrons; simply remove exchange and change the sign of the projectile's charge in the electron-atom scattering codes. However, this procedure yields inaccurate results because, in place of exchange, is the strongly correlated positron-electron configuration of positronium, in the form of virtual states at incident positron energies below the positronium formation threshold ($E_{\text{th}} = E_{\text{ion}}$ - 6.8 eV) and real open channels at energies above the threshold. The accurate representation of positron-electron correlations is essential if accurate values of the scattering parameters are to be obtained.

The other significant difference between the two projectiles is that the positron may annihilate with one of the atomic electrons, yielding two or three gamma-rays depending on the spin state of the annihilating positron-electron pair being singlet or triplet respectively. The spin-averaged cross section for annihilation at incident positron energies below the positronium formation threshold is usually expressed as

$$\sigma_{\text{ann}} = \pi r_0^2 \frac{c}{\nu} Z_{\text{eff}} \ , \tag{1}$$

where r_0 ($= e^2/[4\pi\epsilon m_e c^2]$) is the classical radius of the electron, ν is the speed of the positron and Z_{eff} is an effective number of electrons per atom for annihilation. This parameter is a measure of the probability of the incident positron being at the same position as any one of the atomic electrons, and its value can be determined from the total positron-atom scattering wave function $\Psi(\mathbf{r}_p; \mathbf{r}_1, \mathbf{r}_2, \ldots, \mathbf{r}_Z)$ normalized to unit positron density as the positron coordinate $\mathbf{r}_p \to \infty$ by

$$\begin{aligned} Z_{\text{eff}} &= \sum_{i=1}^{Z} \int \left|\Psi(\mathbf{r}_p; \mathbf{r}_1, \mathbf{r}_2, \ldots, \mathbf{r}_Z)\right|^2 \delta(\mathbf{r}_p - \mathbf{r}_i) d\mathbf{r}_p d\mathbf{r}_1 d\mathbf{r}_2 \ldots d\mathbf{r}_Z \\ &= Z \int \left|\Psi(\mathbf{r}_p; \mathbf{r}_1, \mathbf{r}_2, \ldots, \mathbf{r}_Z)\right|^2 \delta(\mathbf{r}_p - \mathbf{r}_1) d\mathbf{r}_p d\mathbf{r}_1 d\mathbf{r}_2 \ldots d\mathbf{r}_Z \ . \end{aligned} \tag{2}$$

In the Born approximation $Z_{\text{eff}} = Z$, the number of electrons in the target, but the true value of Z_{eff} may be very different owing to distortions of the total wave function arising from the positron-atom interaction. At thermal energies, ~ 40 meV, $Z_{\text{eff}} \approx 9$ for atomic hydrogen, ≈ 4 for helium, and the ratio (Z_{eff}/Z) is several thousand for some organic molecules. The accurate determination of Z_{eff} provides a particularly sensitive test of the accuracy of Ψ, albeit in a very restricted region of configuration space where the positron-electron co-ordinate $r_{1p} \to 0$, because the error in the value of Z_{eff} is of first order in the error in Ψ whereas errors in the scattering parameters, e.g. phase shifts, are usually of second order in the error in Ψ.

The correct behaviour of Ψ close to the coalescent point $r_{1p} \to 0$, where annihilation takes place, is $\Psi \propto (1 - 1/2r_{1p})$ in order to satisfy the 'cusp-condition' that the logarithmic derivative of the total wave function at this point has the value $-1/2$. Thus, if a trial wave function is to provide an accurate representation of the exact wave function, and therefore yield an accurate value for Z_{eff} , it should exhibit a similar linear dependence on the positron-electron coordinate close to the coalescent point.

II. ELASTIC SCATTERING

One of us (JWH) first became involved in detailed theoretical investigations of positron-atom scattering in 1970, soon after the establishment of the experimental positron group at UCL. This group made what were the first reasonably accurate measurements of positron-atom total scattering cross sections, using a mono-energetic positron beam obtained by moderating positrons emitted from the radioactive isotope of sodium, Na^{22} (see Charlton and Humberston, 2001 [4], for more details of experimental procedures). However, such were the uncertainties in the novel procedures employed in these experiments that there was an urgent need for accurate theoretical results to compare with the experimental measurements.

Low energy elastic scattering of positrons by the simplest target, atomic hydrogen, had already been investigated theoretically in considerable detail, most notably by Schwartz [10], but that process was not then amenable to experimental investigation. Therefore, particular theoretical attention was given to low-energy elastic positron- helium scattering, for which experimental results had recently been obtained.

The Kohn variational method was employed for the determination of the scattering phase shifts [7] using trial wave functions of the product form

$$\Psi_t(\mathbf{r}_p;\mathbf{r}_1,\mathbf{r}_2) = F(\mathbf{r}_p;\mathbf{r}_1,\mathbf{r}_2)\Phi_{He}(\mathbf{r}_1,\mathbf{r}_2)\ , \tag{3}$$

where $\Phi_{He}(\mathbf{r}_1,\mathbf{r}_2)$ is the wave function of the helium atom. As Φ_{He} is nodeless, this product form does not impose any restrictions on the possible choice of $F(\mathbf{r}_p;\mathbf{r}_1,\mathbf{r}_2)$, which (for s-wave elastic scattering) was chosen to be

$$\begin{aligned} F(\mathbf{r}_p;\mathbf{r}_1,\mathbf{r}_2) = &\sqrt{\frac{k}{4\pi}}\left\{j_0(kr_p) - \tan\eta\, n_0(kr_p)f(r_p)\right\} \\ &+ (1+P_{12})\left\{\exp\left[-(\alpha r_p + \beta r_1 + \gamma r_2)\right]\sum_i c_i r_p^{k_i} r_1^{l_i} r_{1p}^{m_i} r_2^{n_i} r_{2p}^{p_i} r_{12}^{q_i}\right\}\ , \end{aligned} \tag{4}$$

where $r_{1p} = |\mathbf{r}_1 - \mathbf{r}_p|$ etc., η is the phase shift and c_i are linear variational parameters. Thus, the total trial wave function can be written, in a self-evident nomenclature, as

$$\Psi_T = S + \tan\eta\, C + \sum_i c_i\phi_i \tag{5}$$

The summation included all short-range correlation terms ϕ_i with non-negative integer powers of the six inter-particle distances satisfying the condition

$$k_i + l_i + m_i + n_i + p_i + q_i \le \omega\ , \tag{6}$$

except that q_i, the power of the electron-electron coordinate r_{12}, was limited to even values for numerical convenience. Increasing the integer value of ω then provided a systematic means of improving the trial wave function, allowing the possibility of extrapolation to infinite ω to obtain a very accurate approximation to the exact phase shift. The numbers of short-range terms generated by equation 6 for $\omega = 1, 2, 3, 4$, are 4, 14, 30 and 84 respectively (with q_i even).

The above representation of the short-range correlations is an extension to the four-body system of the correlation functions used so successfully by Schwartz [10] in his studies of positron- and electron-hydrogen scattering.

The target wave function $\Psi_{He}(\mathbf{r}_1, \mathbf{r}_2)$ should be an exact eigenfunction of the target Hamiltonian but, for all targets other than hydrogenic, an approximate function must be used, even though this introduces inconsistencies into the formulation of the Kohn variational method. There is evidence from several other theoretical investigations of positron-atom scattering that this inconsistency has a more serious influence on the results for positrons than for electrons, probably because of the absence of exchange. However, for positron-atom scattering the inconsistencies can be removed by using the 'method of models' [6]. This technique requires the total wave function Ψ to be expressed in the product form of equation 3, and the approximate target wave function is then assumed to be an exact eigenfunction of a model target Hamiltonian. Thus, the real target is replaced by a model target, for which neither the form of its potential energy function nor its energy need be determined. In particular, the electron-electron interaction potential $-1/r_{12}$ is absent from the formulation. The interactions of the incident positron with all the particles in the target, however, retain their exact Coulomb form.

The helium wave function was taken to be [8]

$$\Phi_{He} = \exp\left[-B(r_1 + r_2)\right] \sum_j d_j (r_1 + r_2)^{L_j} (r_1 - r_2)^{M_j} r_{12}^{N_j} , \tag{7}$$

where the summation includes all terms with non-negative integer values of L_j, M_j and N_j satisfying $L_j + M_j + N_j \leq \omega_{He}$, with even M_j to ensure spatial symmetry and even N_j for numerical convenience. The linear parameters d_j were determined using the Rayleigh-Ritz variational method, and the parameter B was varied until the dipole polarizability and the ground-state energy of the target were as close as possible to the exact values.

In order to investigate the sensitivity of the elastic scattering cross sections to the quality of the target wave function, several different helium wave functions were used, from a simple one-term uncorrelated function ($\omega_{He} = 0$) to a fourteen-term function ($\omega_{He} = 4$) providing a very accurate representation of all the relevant properties of the helium atom.

The inclusion of only even powers of r_{12}, the electron-electron coordinate, in the short-range correlation terms and in the helium wave function is, of course, a restriction on the flexibility of the total wave function, but it is probably justifiable because of the repulsion between the two electrons. The advantage of this restriction was that the numerous matrix elements of the form $\langle \phi_i | (H_{tot} - E_{tot}) | \phi_j \rangle$, where H_{tot} and E_{tot} are the total Hamiltonian and the total energy of the system, required in the Kohn variational method could be evaluated exactly. Exact integration over each of the six interparticle coordinates was achieved using Gauss-Legendre quadrature for those coordinates having finite ranges of integration (r_{1p}, r_{2p} and r_{12}) and Gauss-Laguerre quadrature for those having infinite ranges (r_p, r_1 and r_2), with a relatively small number of quadrature points, no more than $(\omega + \omega_{He} + 1)$ in each range of integration. In the evaluation of matrix elements of $(H_{tot} - E_{tot})$ involving the open channel terms S and C, exact integration over all coordinates except r_p was achieved using small numbers of quadrature points, as described above, and the remaining integration over r_p was performed using Gauss-Laguerre quadrature with a sufficiently large number of points to obtain the required accuracy. Further details of these numerical procedures have been given by Armour and Humberston [1].

The regular pattern of convergence of the results with respect to systematic improvements in both the scattering wave function and the target wave function made it possible to conclude that the results obtained using the most elaborate helium wave function were accurate to within a very few per cent of the exact values [7]. Similar calculations for higher partial waves [2], allowed accurate elastic scattering cross sections to be obtained for positron energies up to the lowest inelastic threshold, that for positronium formation at 17.8 eV. The availability of these accurate theoretical results helped to establish which of several sets of measurements of the total cross section was the most accurate, leading to the identification of significant systematic errors, such as the neglect of small-angle scattering, in some experiments.

The Kohn variational method yields the optimum values of the linear variational parameters in the trial wave function Ψ_t, enabling this function to be expressed as an algebraic function of all the interparticle coordinates. It is then a relatively simple task to determine Z_{eff} using equation 2. The value $Z_{\text{eff}} = 3.90$ obtained by Campeanu and Humberston [3] at 40 meV (thermal energy at room temperature) compares well with the experimental result $Z_{\text{eff}} = 3.95$ for thermalized positrons, derived from lifetime experiments. Good agreement was also obtained between theory and experiment for the lifetime spectrum of positrons diffusing and slowing down in dense helium gas before annihilating, the input data to the diffusion equation being the calculated momentum transfer and annihilation cross sections over a range of positron energies [3].

Several other tests of the quality of the positron-helium scattering wave function at the positron-electron coalescent point have been made. The most impressive of these is the comparison between theoretical and experimental values of the distribution function for the Doppler broadening of the two gamma rays arising from the annihilation of singlet positron-electron pairs. The Doppler shift in the energy of a gamma-ray from its value of $m_e c^2 = 511$ keV is a measure of the total momentum **p** of the positron-electron pair at the coalescent point immediately prior to annihilation, and the relevant probability distribution function is

$$\Gamma(p) \propto \int \left| \exp(-i\mathbf{p}\mathbf{r}_1)\Psi(\mathbf{r}_p;\mathbf{r}_1,\mathbf{r}_2)\right|^2 d\mathbf{r}_p d\mathbf{r}_1 d\mathbf{r}_2 \, . \tag{8}$$

Excellent agreement was obtained between the most accurate theoretical results and the most reliable measurements [16].

III. POSITRONIUM FORMATION

The accurate determination of the cross section for the rearrangement process of positronium formation,

$$e^+ + A \rightarrow Ps + A^+ \, , \tag{9}$$

provides an even more demanding test of the quality of the representation of positron-electron correlations owing to the extended structure of the total wave function as the two fragments, positronium and the ion, form in the rearrangement channel. Indeed, this structure begins to manifest itself quite significantly at energies just below the positronium formation threshold, to the extent that additional correlation functions providing a more explicit representation of the configuration of positronium weakly bound to He^+ needed to be included in the trial wave function for well-converged results to be obtained and threshold features revealed [13].

The most detailed theoretical investigations of positronium formation in positron-helium scattering have been made by Van Reeth and Humberston [15,14,13] in the Ore gap, the energy range between the positronium formation threshold at 17.8 eV and the lowest excitation threshold of helium (which for positron impact is the 2^1S_0 state, with its threshold at 20.6 eV, and not the 2^3S_1 state with its threshold at 19.8 eV). These authors employed a two-channel version of the Kohn variational method for the **K**-matrix, with trial wave functions containing many short-range correlation functions of similar form to those described above.

The 'method of models' cannot be used in a rearrangement process, and inconsistencies in the formulation of the scattering problem associated with the use of an inexact target wave function are therefore inevitable, resulting in a breakdown in convergence of the scattering data when simple helium wave functions are used. In order to minimize the possible effect of these inconsistencies, and avoid problems of convergence, a very accurate 22-term helium wave function was used ($\omega_{He} = 4$, but with no restriction to even powers of r_{12}) [15].

Matrix elements of the electron-electron interaction potential $1/r_{12}$ must now be evaluated numerically but exact integration of matrix elements of $(H_{tot} - E_{tot})$ can no longer be achieved, even between two short-range correlation terms. Further significant numerical complications arise from the presence of the coordinate of the centre of mass of the positronium relative to the helium nucleus $\rho = |(r_p + r_1)|/2$, and its electron-exchange form $\rho' = |(r_p + r_2)|/2$, in matrix elements of $(H_{tot} - E_{tot})$ involving open positronium terms in the trial wave function. Also, the limits of integration for the r_{1p} and r_{2p} variables introduce cusps at $r_1 = r_p$ in the r_1 integrand and $r_2 = r_p$ in the r_2 integrand for some matrix elements. Numerical procedures were developed whereby the overall range of integration for each of these variables was split into two regions, one on either side of the cusp, with Gauss-Legendre quadrature used below the cusp and Gauss-Laguerre quadrature above.

These numerical procedures, when used with very flexible trial wave functions containing as many as 502 short-range correlation functions ($\omega = 6$ with no restriction to even powers of r_{12}), enabled very accurate cross sections to be obtained for both elastic scattering and positronium formation that exhibited interesting structures very close to the positronium formation threshold [13].

More recently, one of us (PVR) has developed a new approach to the integration of the matrix elements containing only short-range terms. The method is based on the work of Drake and Yan [5,17] for bound states of lithium, in which the inter-electron coordinates are expanded in Legendre polynomials using a formula developed by Perkins [9]. This expansion allows the integration over the coordinates between the leptons and the nucleus to be carried out using an analytical expression, which is mainly an expansion in terms of hypergeometric functions ${}_2F_1$. The method has been applied to positron-helium scattering and to positronium scattering by atomic hydrogen using wave functions containing up to 918 terms [12,11] and the convergence of the integration procedure was found to be very good, even for odd and negative powers of r_{ij}. Also, this procedure enabled the evaluation of the various terms that make up each matrix element to be reorganized, resulting in a very significant improvement in both the running time of the code and the precision of the calculations.

Further improvements in the numerical techniques, together with increases in computer power, should allow similarly detailed investigations to be made of positron-helium scattering at energies above the Ore gap where other inelastic processes occur, and even of five-body systems such as positron-lithium scattering and positronium-helium scattering.

[1] Armour, E. A. G. and Humberston, J.W., 1991, *Phys. Reports* **204** 165
[2] Campeanu, R.I. and Humberston, J.W., 1975, *J. Phys. B* **8** L244
[3] Campeanu, R.I. and Humberston, J.W., 1977, *J. Phys. B* **10** L153
[4] Charlton, M. and Humberston, J.W., 2001, *Positron Physics*, Cambridge University Press
[5] Drake, G.W.F. and Yan, Z-C., 1995, *Phys. Rev. A* **52** 3681
[6] Drachman, R.J., 1972, *J. Phys. B* **5** L30
[7] Humberston, J.W., 1973, *J. Phys. B* **6** L305
[8] Hylleraas, E.A., 1929, *Z. f. Physik* **54** 347
[9] Perkins, J.F., 1968, *J. Chem. Phys.* **48** 1985
[10] Schwartz, C., 1961, *Phys. Rev.* **124** 1468
[11] Van Reeth, P. and Humberston, J.W., 2004, *Nucl. Instr. Meth. B* **221** 140
[12] Van Reeth, P. and Humberston, J.W., 2003, *J. Phys. B* **36** 1923
[13] Van Reeth, P. and Humberston, J.W., 1999, *J. Phys. B* **32** 3651
[14] Van Reeth, P. and Humberston, J.W., 1997, *J. Phys. B* **30** L95
[15] Van Reeth, P. and Humberston, J.W., 1995, *J. Phys. B* **28** L23
[16] Van Reeth, P., Humberston, J.W., Iwata, K., Greaves, R.G. and Surko, C.M., 1996, *J. Phys. B.* **29** L465
[17] Yan, Z-C and Drake, G.W.F., 1997, *J. Phys. B* **30** 4723

E. A. G. Armour, J. Franz and J. Tennyson (eds.)
Explicitly Correlated Wavefunctions

Calculations of Bound States and Resonance States of Few-Body Atomic Systems with Screened Coulomb Potentials

Y. K. Ho and Sabyasachi Kar

Institute of Atomic and Molecular Sciences, Academia Sinica, P.O. Box 23-166, Taipei, Taiwan 106, R.O.C.

The influence of plasma on the bound states and resonance states of few-body atomic systems has been invesigated recent studies. The plasma effect has been considered by using the screened Coulomb potential obtained from the Debye model. Our recent reults for the ground states and the resonance states of Ps^- ion and He atom in plasmas for various screening parameters along with the bound excited $S-$, $P-$, $D-$ states of He are discussed.

I. INTRODUCTION

The study of the effect of screened Coulomb potential representing dense plasmas on the few-body atomic systems is an interesting and relevant topic of reseach in recent years. Effect of plasma on an embedded atom/ion is to produce an effective screening, characterized by a fundamental length scale determined by the temperature (T) and number density (n) of the charge particles, on the Coulomb potential. Several studies have been performed in recent years on the few-body atomic systems embedded in weakly coupled plasmas ([1-8] and references therein) considering the Debye screening model. It is well known in the literature that the Debye-Hückel model is a good approximation for certain specific conditions, for example, in the low-density and high temperature limit. Also the concept of Debye-Hückel screening model has been derived from the weakly coupled plasmas in which coupling constants are much smaller than 1. There are several applications of weakly coupled plasmas [9], as for example, $n \approx 10^{11} cm^{-3}$, $T \approx 10^4 K$ for gaseous discharge plasma, $n \approx 10^{16} cm^{-3}$, $T \approx 10^8 K$ for plasma in controlled thermonuclear experiments and $n \approx 10^6 cm^{-3}$, $T \approx 10^8 K$ for plasma in soloar corona. For those plasmas coupling constants are 10^{-3}, 10^{-5}, and 10^{-7} respectively. The low-density and high-temperature plasma has sevral applications in the intersteller space, for example, for dilute stars having electron densities range of 10^{-2}-$10^{0} cm^{-3}$ and the temperature $\approx 10^{4} K$.

In our recent works, we have made theoretical investigations on the bound states and the resonance states of the Ps^- ion and the He atom using the Debye screening model. A highly correlated wave function has been used to repesent the correlation effect between the charge particles. We have obtained the ground and bound excited states energies in the framework of Rayleigh-Ritz variation principle using a quasi-random process [10]. For resonance states, the density of resonance states has been calculated using the stabilization method proposed by Mamdelshatm *et al* [11]. All the calculations have been performed using quadruple precision arithmetic (32 significant figures) on DEC-ALPHA machines in the UNIX environments. Atomic units have been used throughout the investigations.

II. CALCULATIONS

In the Debye-Hückel model for plasmas, the interaction between two charges a and b is represented by the Yukawa-type potential: $V(r_a, r_b) = q_a q_b exp(-\mu|r_a - r_b|/|r_a - r_b|)$. The screening parameter μ (the Debye length, $\lambda_D = 1/\mu$) is proportional to the square root of the ratio of number density (n) and temperature (T) of the charge particles in the plasma. For different choice of n and T, μ represents a set of plasma conditions. The Debye prameter μ can charaterize both the hot-dense plasma and low-density warm plasma as μ is scaled by $\sqrt{n/T}$ [8]. For Ps^- and He, the non-relativistic Hamiltonian for a system (e^-, e^-, Z^+) is given by

$$H = -\frac{1}{2}\sum_{i=1}^{3}\frac{1}{m_i}\nabla_i^2 + \sum_{i<j=1}^{3} V(r_j, r_i), \tag{1}$$

where q_i ($i = 1, 2, 3$) are the charges of the three particles and m_i ($i = 1, 2, 3$) are their respective masses. For Ps^- ion, $q_1 = q_2 = -1$, $q_3 = 1$ and $m_1 = m_2 = m_3 = 1$. For He atom, $q_3 = 2$, $m_1 = m_2 = 1$ and the third particle is of infinite mass.

We have employed the wave functions

$$\Psi = (1 \pm P_{12})\sum_{i=1}^{N} r_{31}^{L} P_L(\cos\theta_1) exp[(-\alpha_i r_{32} - \beta_i r_{31} - \gamma_i r_{21})\omega], \tag{2}$$

where L =0 for the $S-$states, $L = 1, 2$ for $P-$, $D-$states, '+' for singlet, '−' stands for triplet states, α_i, β_i, γ_i are the non-linear variation parameters, P_L

denontes the Legendre polynomial of order L, P_{12} is the permutation operator on the subscripts 1 and 2 representing the electrons. The scaling factor ω acts as the reciprocal range of a 'soft-wall' for resonance calculations whereas ω is set to 1 for ground state calculations.

We have performed accurate variational calculations in the framework of Rayleigh-Ritz variational principle to obtain the ground state enegies (the only bound state) of Ps^- ion and the bound excited $^{1,3}S-$, $^{1,3}P-$, $^{1,3}D-$ state energies of the He atom in plasmas for various Debye lengths using a quasi-random process [10]. According to the quasi-random process [10], the parameters α_i, β_i and γ_i are chosen from the three positive interval $[A_1^{(k)}, A_2^{(k)}]$, $[B_1^{(k)}, B_2^{(k)}]$ and $[C_1^{(k)}, C_2^{(k)}]$; where $k = mod(i, 3) + 1$, $1 \le i \le N$.

$$\begin{aligned} \alpha_i &= \eta_1^{(k)}[<< \frac{1}{2}i(i+1)\sqrt{2} >> (A_2^{(k)} - A_1^{(k)}) + A_1^{(k)}] \\ \beta_i &= \eta_2^{(k)}[<< \frac{1}{2}i(i+1)\sqrt{3} >> (B_2^{(k)} - B_1^{(k)}) + B_1^{(k)}], \\ \gamma_i &= \eta_3^{(k)}[<< \frac{1}{2}i(i+1)\sqrt{5} >> (C_2^{(k)} - C_1^{(k)}) + C_1^{(k)}] \end{aligned} \tag{3}$$

where the symbol $<< ... >>$ designates the fractional part of a real number. The positive scaling factors $\eta_1^{(k)}$, $\eta_2^{(k)}$ and $\eta_3^{(k)}$ are set to have values equal to 1 in the first stage and in the second stage they are varied. But for our recent investigation, it has been observed that the better optimization can be obtained by selecting $A_1^{(k)} = 0$, $A_2^{(k)} = a$; $B_1^{(k)} = 0$, $B_2^{(k)} = b$; $C_1^{(k)} = 0$, $C_2^{(k)} = c$ and $\eta_1^{(k)} = 1$, $\eta_2^{(k)} = 1$, $\eta_3^{(k)} = \eta$. Ultimately, the four variation parameters a, b, c and η are used in the entire calculations. The necessary integrals involving the screened Coulomb potentials with basis functions (2) have been discussed in the literature ([1], references therein). We have calculated the energy levels by diagonalizing the Hamiltonian (1) with the wave functions (2) following the quasi-random process for each spin staes, for each $S-$, $P-$, $D-$ states, and for each Debye lengths.

The parameters a, b, c and η are chosen so that the ground and bound excited states energies are minimum for different shielding parameters using certain basis sets. For resonance calculations, we next let ω having values other than $\omega = 1$. The variational parameters ω in the wave function (2) are acting as the reciprocal range of a 'soft' wall ([4-6, 11]). Varying ω from 0.3-1.0, we have computed the energy levels $E(\omega)$ for each Debye length. We then construct the stabilization diagrams (as shown in figures 1(a) and 1(b)) by plotting $E(\omega)$ versus ω. If there is a resonance at energy E, the stabilized or slowly varying energy levels will appear in the stabilization plateau. The details successful applications of this simple and powerful method are available in the works of Ho and co-workers [4-6] and references therein. Varying the Debye length from infinity to small values, the resonance energies and widths have been obtained for different resonance states.

To extract the resonance energy E_r and the resonance width Γ, we have calculated the density of resonance states for a single energy level with the help of the following formula,

$$\rho_n(E) = \left| \frac{E_n(\omega_{i+1}) - E_n(\omega_{i-1})}{\omega_{i+1} - \omega_{i-1}} \right|^{-1}_{E_n(\omega_i)=E}, \tag{4}$$

where the index i is the i-th value for ω and the index n is for the n-th resonance. After calculating the density of resonance states $\rho_n(E)$ with the above formula (4), we fit it to the following Lorentzian form that yields resonance energy E_r and a total width Γ, with

$$\rho_n(E) = y_0 + \frac{A}{\pi} \frac{\Gamma/2}{(E - E_r)^2 + (\Gamma/2)^2}, \tag{5}$$

where y_0 is the baseline offset, A is the total area under the curve, E_r is the center of the peak, and Γ denotes the full width of the peak of the curve at half height.

III. RESULTS OF PLASMA-EMBEDDED Ps^-

The positronium negative ion has been first observed by Mills [12], and he subsequently measured its annihilation rate [13]. Recent experimental developments in positron trapping and accumulation technology [14-17] have opened up access to study various properties of such a three-lepton system. With the recent development of the modern positron sources, it is of great inportance to invesigate the ground and resonance states of Ps^- under the influence of plasma screening.

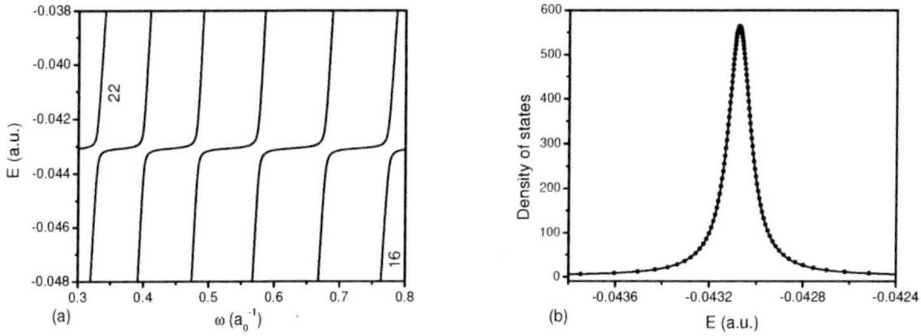

FIG. 1. Stabilization plots for the lowest $^3P^o$ state of Ps^- in plasmas for $\lambda_D = 30$. (b) Calculated density (circles) and the best fitted Lorentzian form (solid line) corresponding to the $^3P^o$ state of Ps^- for the 18th energy level in (a).

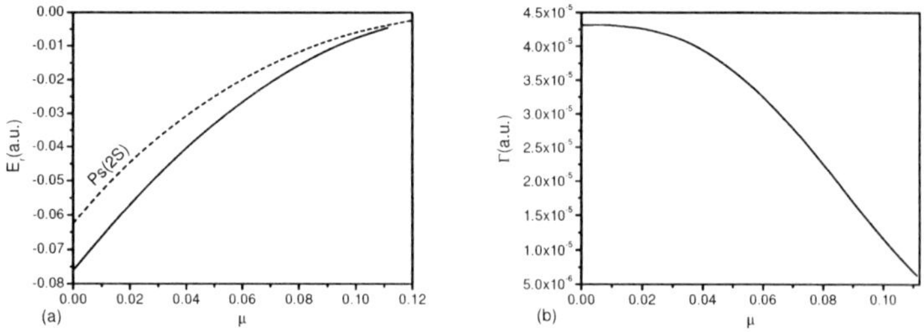

FIG. 2. The $2s^2\ ^1S^e$ resonance energies in (a) and the corresponding resonance widths in (b) of Ps^- in plasmas in terms of Debye parameter μ.

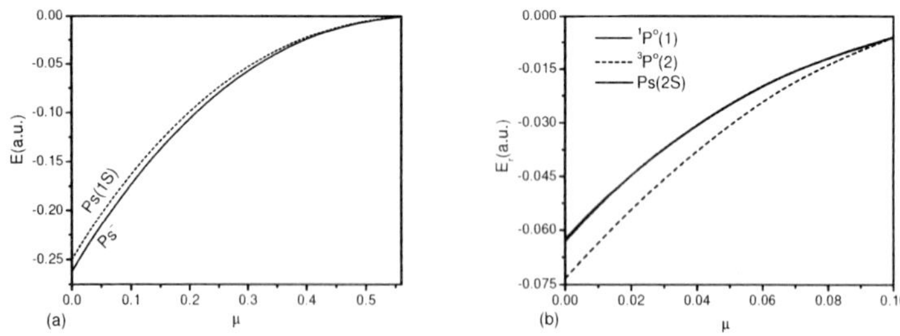

FIG. 3. The ground-state energies in (a) and the lowest $^{1,3}P^o$ resonance energies in (b) of Ps^- as a function of Debye screening parameter μ.

Following the method described above we have investigated the bound ^1S state and for resonances the $2s^2\ ^1S^e$ state, one $^1P^o$ state, and two $^3P^o$ states of Ps^- in desnse plasmas for different Debye lengths. Figure 1(a) shows the stabilization diagram for the Debye length $\lambda_D = 30$ and the fitting of the calculated density of states(figure 1(b)) corresponding to the 18th energy levels in figure 1(a) which exibits the best fit with the least χ^2 and the square of the correlation coefficient very close to 1. From the fit, we have obtained the resonace energy, $E_r = -0.033172$ a.u. and the corresponding resonance width $\Gamma = 3.64 \times 10^{-5}$ a.u.. Similarly, for each Debye length, we have constructed the stabilization plot and search for the best fittings from the each energy levels in the stabilization diagams by calculating the desity of states using the formula (4). We have presented the $2s^2\ ^1S^e$ resoance energies and the corresponding widths in figure 2 for various sheilding parameters. The gound state energies (see also table 1) and the lowest $^{1,3}P^o$ resonance energies in terms of μ are presented in figure 3(a) and 3(b) respectively. The $^3P^o(2)$ resonance energies are very close to the

$^1P^o(1)$ resonace energies for each Debye lengths and so we have not included the $^3P^o(2)$ resonance energies in figure 3(b). The resonace width for the three P-states are shown in figure 4. From figure 3(a), it is apparent that the ground state energy approches the Ps(1S) threshold energies with increase of plasma strength and both the systems will be ionized near $\lambda_D = 1.8$. Figures 2(a) and 3(a) show that the lowest $^1S^e$, $^3P^o$ resonace energies approche the Ps(2S) threshold with the increase of plasma strengths. Also it is evident from figures 2(b)and 4(a)-(c), that the $^1P^o$ resonance widths increase whereas the other S-, P- states resonance widths decrase with increasing plasma strength. Detailed discussion can be found in Ref.[6].

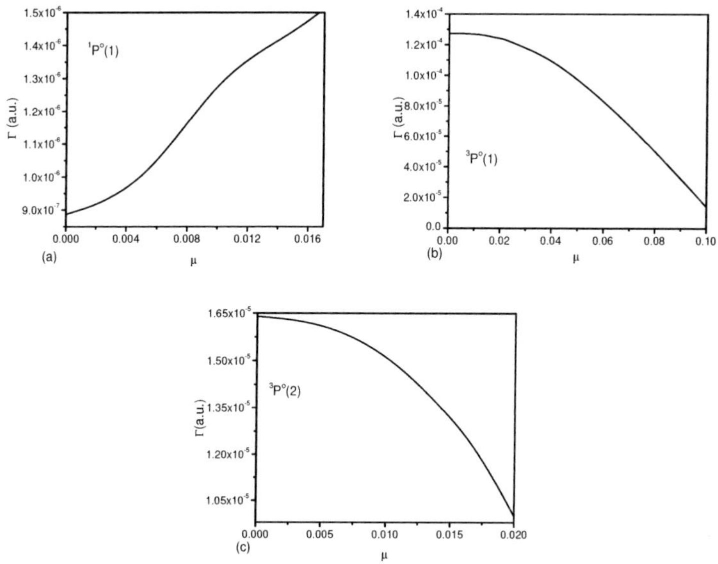

FIG. 4. The resonace widths for $^1P^o(1)$ in (a), $^3P^o(1)$ in (b) and $^3P^o(2)$ in (c) states of Ps$^-$ as a function of Debye screening parameter μ.

IV. RESULTS OF PLASMA-EMBEDDED He

Following the technique discussed above we have calculated the bound $^{1,3}S$, $^{1,3}P$, $^{1,3}D$ and the $^{1,3}P^o$ resonance states of helium atom embedded in Debye plasmas. Recent experimental investigations [18, 19] on the doubly excited P-states of helium in the un-screened case, the investigation on the bound excited and doubly excited states under the influence of Debye screening has become an important subject. We have presented the bound excited $^{1,3}S$ states and the excitation energies of few lowest singlet states in figure 5. The ground (see Table 1) and bound excited states energeis are very close to the $He^+(1S)$ threshold energies with the increase of the screening parameter μ. Figure 5(a) shows that higher lying states approach the ionization limit one after another with the increase of plasma strength. Deatil results on the ground and bound excited energies are available in our earlier works [5]. To calculate the $^{1,3}P^o$ resonance states of He in plasmas for different Debye lengths, we have constructed the stabilization plots (see figure 6 for example) for each Debye lengths. After indentifing the resonance positions for the $2snp_+(2 \leq n \leq 5)$, $2snp_-(3 \leq n \leq 5)$, $2pnd(n = 3, 4)$ $^{1,3}P^o$ states and following the method discussed above, we have obtained a total of 9 resonances below the $He^+(2S)$ threshold for various Debye lengths. We have presented the wavelengths for the photoabsorption of $^1P^o$ resonaces and widths in figure 7 and 8. It is clear from figure 8, the resonance width decreases with increasing of plasma strength. Deatails results and discusstions on the doubly-excited $S-$, $P-$ states of He are found in our recent works [5].

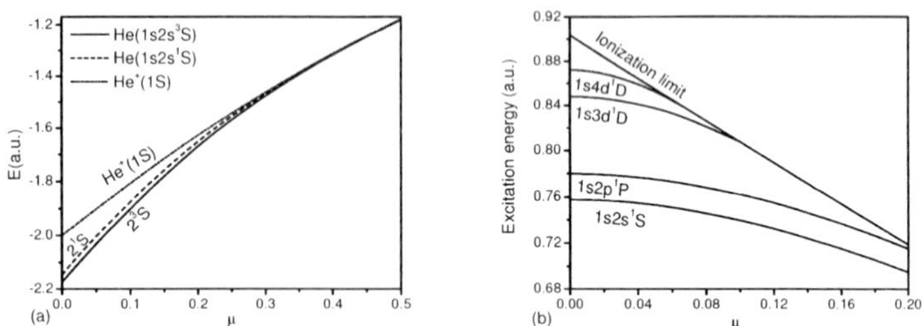

FIG. 5. The bound excited S-states and few lowest singlet excitation energies of He in plasmas for various Debye lengths.

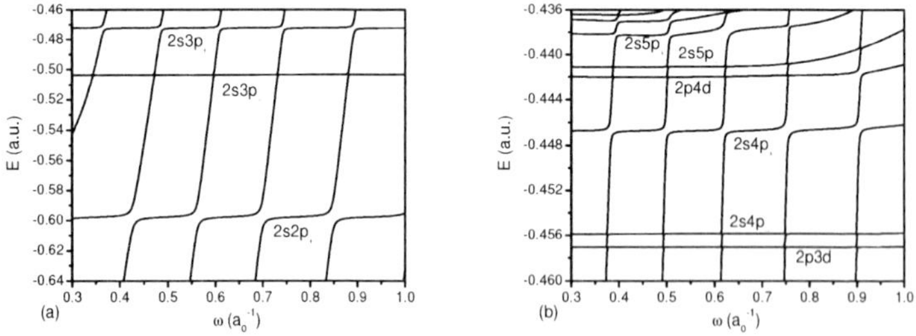

FIG. 6. Stabilization plots for the $2s2p_+$, $2s3p_-$, $2s3p_+$ in (a) and the $2p3d$, $2s4p_-$, $2s4p_+$, $2p4d$, $2s5p_-$, $2s5p_+$ $^1P^o$ states of He embedded in Debye plasma for $\lambda_D = 30$.

V. CONCLUSIONS

We have reported the influence of dense plasma on the bound and resonace states of some few-body atomic systems (Ps^- and He) using the Debye screening concept of plasmas modelling and employing highly correlated wave fuctions. To extract resonance parameters, we have used the stabilization method [4-6, 11] which is a simple and powerful method that needs only the bound state type wave functions. The resonance states calculations for the few-body systems in dense plasmas [4-6] were the first investigations in the literature. It is imortant to mention here that Debye screening is widely used in plasma modelling for its simplicity and admits sets of plasma conditions. However a shortcoming of the Debye model is its limitation to static screening. Correction to dynamic screening is very important for the strongly coupled plasma. However, corrections due to dynamic screening have not considered in our reported results as we have restricted ourself in the study of few-body systems in weakly coupled plasmas. Results reported are useful to atomic physics, plasma physics and astrophysics research communities.

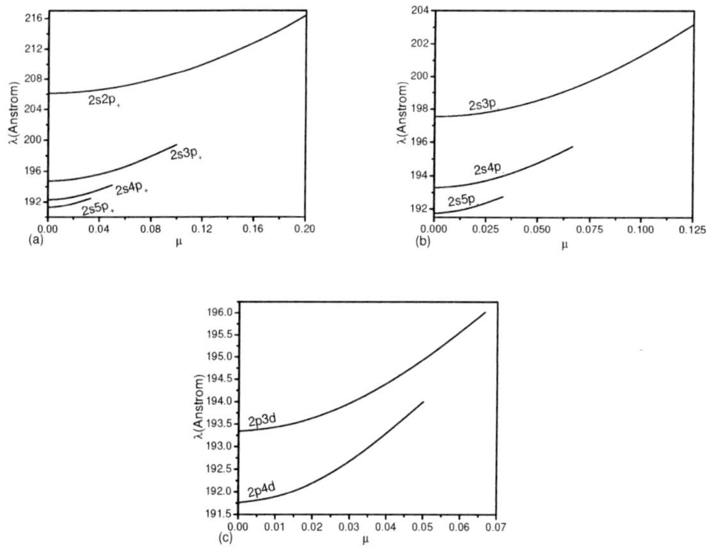

FIG. 7. The wavelengths for the absorption resonances for the $2snp_+ \, {}^1P^o$ in (a), $2snp_- \, {}^1P^o$ in (b), $2pnd \, {}^1P^o$ in (c) states in helium for various Debye parameters.

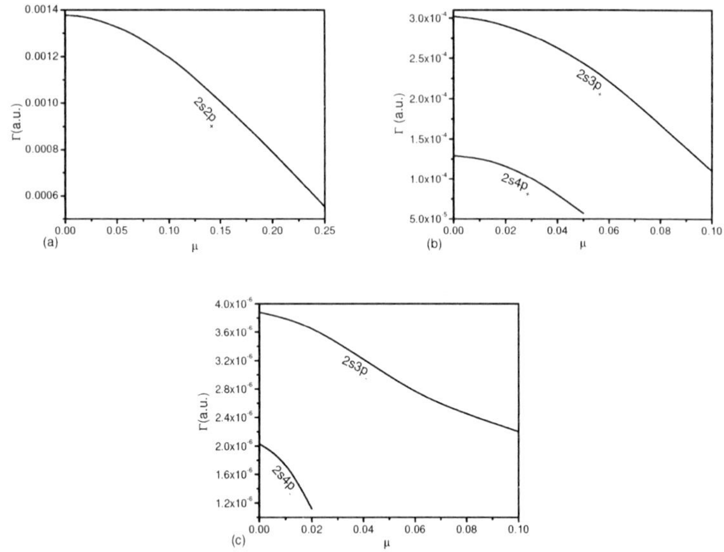

FIG. 8. The resonance widths for $2snp_+ \ ^1P^o$ in (a) and (b), $2snp_- \ ^1P^o$ in (c), states in the helium atom for various Debye parameters.

λ_D	Ground state of Ps^-	λ_D	Ground state of He
∞	$-0.2620050702325\ ^a$	∞	-2.90372437700^a
	-0.26200507023298^b		-2.90372437703^b
50	-0.2424865663745^a	50	-2.84418057552^a
10	-0.173618160010^a	10	-2.61485294693^a
5	-0.1064096776153^a	5	-2.34700618425^a
1.8	-0.0011185^a	0.45	-0.0081^a

[a] Our recent calculations[5, 6], [b] Best results(references in [5, 6])

Table 1. The ground state ^1S-state energies of Ps^- and He atom under Debye screening

Acknowledgement

The work is supported by the National Science Council of Taiwan, R. O. C.

[1] S. T. Dai and P. Winkler, Phys. Rev. A **52**, 216 (2001)

[2] I. Silances, J. M. Mercero and J. M. Ugalde, Phys. Rev. E **66** , 026408 (2002).

[3] B. Saha, T. K. Mukherjee, P. K. Mukherjee, Chem. Phys. Lett. **373** (2003) 218.

[4] S. Kar and Y. K. Ho, Phys. Rev. E **70**, 066411; New J. Phys. **7**, 141(2005); J. Phys. B **38**, 3299 (2005)

[5] S. Kar and Y. K. Ho, Chem. Phys. Lett. **402**, 544 (2005);Phys. Rev. A **72**, 010703 (2005); Int. J. Quant. Chem. **106**, 814 (2006);to be published (2006).

[6] S. Kar and Y. K. Ho, Phys. Rev. A **71**, 052503 (2005);Phys. Rev. A **73**, 032502(2006); to be published (2006).

[7] A. N. Sil and P. K. Mukherjee, Int. J. Quant. Chem. **102**, 1061 (2005).

[8] H. Okutsu, T. Sako, K. Yamanouchi, G. H. F. Diercksen, J. Phys. B **38**, 917 (2005).

[9] Ichimaru, S., in Plasma Physics, The Benjamin/Cummings Publishing Company, Inc.: (Menlo Park, California, 1986) pp5-10.

[10] A. M. Frolov, Phys. Rev. E **64**, 036704 (2001).

[11] V. A. Mandelshtam, T. R. Ravuri and H. S. Taylor, Phys. Rev. Lett. **70**, 1932 (1993).

[12] A. P. Mills, Jr., Phys. Rev. Lett. **46**, 717 (1981).

[13] A. P. Mills, Jr., Phys. Rev. Lett. **50**, 671 (1983).

[14] D. Schwalm *et al.*, Nucl. Instr. Meth. **221**, 185 (2004).

[15] P. Balling *et al.*, Nucl. Instr. Meth. **221**, 200 (2004).

[16] F. Fleiser *et al.*, Cann. J. Phys. **83**, 413 (2005).

[17] D. B. Cassidy *et al.*, Phys. Rev. Lett. **83**, 413 (2005).

[18] R. P. Madden and K. Codling, K. Ap. J. **141**, 364 (1965); Phys. Rev. Lett. **10**, 516 (1963).

[19] Domke *et al.*, Phys. Rev. Lett. **66**, 1306 (1991); **69**, 1171 (1992); Phys. Rev. A **53**, 1424 (1996).

E. A. G. Armour, J. Franz and J. Tennyson (eds.)
Explicitly Correlated Wavefunctions

Treatment of positron-electron correlation using Hylleraas-type functions

Edward A. G. Armour

School of Mathematical Sciences, University of Nottingham, Nottingham NG7 2RD, UK

Hylleraas-type functions are functions that contain the electron-electron or positron-electron distance as a linear factor. In this article, I begin by describing how basis sets containing these functions have been used in variational calculations to obtain very accurate results for ground-state energies of simple atoms and molecules. I go on to describe how they have been used to obtain accurate results for low-energy positron-H_2 elastic scattering, using the Kohn variational method. I conclude by indicating how recent calculations on the interaction between helium-antihydrogen, that make use of Hylleraas-type functions, will be adapted to treat positron-H_2 scattering with inclusion of the positronium formation channel.

I. INTRODUCTION

The Norwegian physicist Hylleraas carried out a variational calculation of the ground-state energy of helium in 1929 in which he included basis functions which contained the interelectron distance as a linear factor [1]. His basis functions were of the form

$$\chi_i = s^{a_i} t^{b_i} u^{c_i} e^{-\frac{1}{2}\alpha s} \tag{1}$$

where α is a parameter, a_i, b_i and c_i are non-negative integers and b_i is even,

$$s = r_1 + r_2, \quad t = r_2 - r_1, \quad u = r_{12} = |\boldsymbol{r}_2 - \boldsymbol{r}_1|,$$

where $\boldsymbol{r}_1$ and $\boldsymbol{r}_2$ are the position vectors of electrons relative to the nucleus. Much improved accuracy of the energy was obtained by the inclusion of basis functions with $c_i = 1$. These are '*Hylleraas-type functions*'.

Hylleraas was able to include r_{12} as one of the coordinates in the integration. The contributions in these proceedings from Humberston and Van Reeth and

Ho and Kar show how this method can be extended to make possible, for example, calculations of positron scattering by helium both below and above the positronium formation threshold.

In this article, I will show how this method was applied to positron scattering by a hydrogen molecule below the positronium formation threshold. I will go on to describe how the method has been applied to calculations of the interaction of antihydrogen with hydrogen and helium atoms. I will conclude by outlining how the antihydrogen-helium calculations will be adapted to make possible a calculation of positron-H_2 scattering above the positronium formation threshold.

II. APPLICATION OF THE HYLLERAAS METHOD TO MOLECULES

Only four years after the above calculation on helium, James and Coolidge [2] carried out a calculation of the ground-state electronic energy of the hydrogen molecule using a basis set that included Hylleraas-type functions. Hylleraas had been able to use spherical polar coordinates in his calculation on helium and also to include r_{12} as a variable of integration. The natural analogue of spherical polar coordinates for a diatomic molecule are prolate spheroidal coordinates. See, for example, ref. [3].

These are an orthogonal system of coordinates (λ, μ, ϕ). If P is any point and r_A and r_B are the distances of P from the nuclei A and B, respectively, then

$$\lambda = \frac{r_A + r_B}{R}, \quad \mu = \frac{r_A - r_B}{R},$$

where R is the distance between the nuclei. If we take Cartesian coordinates with origin midway between the nuclei and z-axis along the internuclear axis in the direction A to B and arbitrarily chosen x and y-axes, ϕ is the usual azimuthal angle of spherical polar coordinates.

The basis set used by James and Coolidge was of the form

$$\chi_i = \frac{1}{2\pi}(\lambda_1^{a_i}\lambda_2^{b_i}\mu_1^{c_i}\mu_2^{d_i} + \lambda_1^{b_i}\lambda_2^{a_i}\mu_1^{d_i}\mu_2^{c_i})\rho_{12}^{e_i}\exp[-\alpha(\lambda_1 + \lambda_2)], \tag{2}$$

where α is a parameter, a_i, b_i, c_i, d_i and e_i are non-negative integers and

$$\rho_{12} = \frac{2}{R} r_{12}. \tag{3}$$

As it was not possible to include r_{12} as a variable of integration, James and Coolidge used the Neumann expansion [4]

$$\frac{1}{\rho_{12}} = \sum_{\tau=0}^{\infty}\sum_{\nu=0}^{\tau} D_\tau^\nu P_\tau^\nu(\lambda_<)Q_\tau^\nu(\lambda_>)P_\tau^\nu(\mu_1)P_\tau^\nu(\mu_2)\cos[\nu(\phi_1 - \phi_2)], \tag{4}$$

where D_τ^ν is a coefficient, $\lambda_<$ and $\lambda_>$ are the lesser and greater, respectively, of λ_1 and λ_2 and P_τ^ν and Q_τ^ν are the first and second solutions, respectively, to the Legendre equation or, if $\nu \neq 0$, the associated Legendre equation.

It is easy to show using equation (3) and the expressions for x, y and z in terms of λ, μ and ϕ [3] that

$$\rho_{12}^2 = \lambda_1^2 + \lambda_2^2 + \mu_1^2 + \mu_2^2 - 2 - 2\lambda_1\lambda_2\mu_1\mu_2 - 2M_{12}\cos(\phi_1 - \phi_2), \qquad (5)$$

where

$$M_{12} = [(\lambda_1^2 - 1)(1 - \mu_1^2)(\lambda_2^2 - 1)(1 - \mu_2^2)]^{\frac{1}{2}} \qquad (6)$$

$e_i \leq 2$ in the basis functions (2). Thus odd powers of ρ_{12}^p $(p = -1, 1, 3)$ occurred in matrix elements required for the variational calculation. These were evaluated using the Neumann expansion (4) and, where necessary, equation (5).

As the calculation was carried out only for the region around the equilibrium value $R = 1.4a_0$ and not for the region in which H_2 begins to separate up into $H + H$, it was not necessary to include μ_1 and μ_2 in the exponent in the basis functions. Consequently all the series expansions involving $\dfrac{1}{\rho_{12}}$ in the matrix element evaluations terminated after a finite number of terms. All the resulting integrals were evaluated analytically.

The calculations were carried out on hand calculating machines and took approximately six months. James and Coolidge obtained a binding energy of 4.697 eV using a thirteen term trial function containing 4 Hylleraas-type functions. The experimental value is 4.747 eV.

With the advent of computers, Kołos and Roothaan [5] were able to obtain excellent agreement for the binding energy using a 50 term trial function containing 14 Hylleraas-type functions. Together with Wolniewicz, Kołos went on to make a wide application of the method used in ref. [5] to states of two electron diatomic molecules and ions. See, for example, ref. [6].

It is a very big step to extend these calculations to molecules containing even three electrons. In pioneering research in 1977, Clary [7] carried out calculations on He_2^+ and He_2 using basis sets that contained Hylleraas-type functions with an r_{ij} factor in each term. This required the evaluation of integrals involving, for example,

$$\frac{r_{12}}{r_{13}}, \quad r_{12}r_{13} \quad \text{and} \quad \frac{r_{12}r_{13}}{r_{23}}.$$

To allow for the separation of He_2^+ into $He + He^+$, it was necessary to take the exponential factors in the basis to be of the form

$$\exp\left[\sum_i(-A_i\lambda_i + B_i\mu_i)\right]$$

where $B_i \neq 0$.

Double Neumann expansions are required for $\frac{r_{12}}{r_{13}}$ and $r_{12}r_{13}$ and triple for $\frac{r_{12}r_{13}}{r_{23}}$. As $B_i \neq 0$, none of the expansions terminate, but convergence is rapid for small $|B_i|$. Despite the complexities of the expansions, Clary was able to carry out all the required evaluations using no more than two dimensional numerical integration by adapting and correcting formulae devised by Rothstein [8], based on (4).

Clary obtained 90% of the correlation energy of He_2^+ at $R = 2.065a_0$ using 29 basis functions, of which 3 were Hylleraas-type functions. His results for He_2 at $R = 1.0a_0$ were similar.

III. CALCULATIONS ON e^+ H_2 SCATTERING USING HYLLERAAS-TYPE FUNCTIONS

Prolate spheroidal coordinates were first used in scattering calculations by Massey and Ridley [9] for eH_2 and Massey and Moussa [10] for e^+H_2. Use of these coordinates makes it possible to carry out calculations for e^+H_2 using the Kohn variational method similar to those carried out by Humberston and Van Reeth for e^+H and e^+He scattering [11,12].

The Kohn variational method is the analogue for scattering wave functions of the Rayleigh Ritz variational method for bound state wave functions. If $\hat{H}$ is the Hamiltonian of the system under consideration and Ψ_t is a trial function, it is not possible to evaluate $\langle\Psi_t|\hat{H}|\Psi_t\rangle$ if Ψ_t is a scattering wave function, as it is not square integrable. However, it is possible to evaluate $\langle\Psi_t|\hat{H} - E|\Psi_t\rangle$ if E is the energy of the scattering state under consideration and Ψ_t has the same general asymptotic form as the exact eigenfunction. A closely related functional is made stationary in the Kohn method.

With Martin Plummer and David Baker [13], I carried out calculations of low-energy elastic scattering of a positron by H_2 using a trial function which was of the form

$$\Psi_t = \left[S + a_t Cp + \sum_{j=1}^{N} b_j \phi_j \right] \Psi_G, \tag{7}$$

if mixing of spheroidal partial waves is neglected. In the atomic case, S and C would be the regular and irregular s-wave solutions, respectively, to the free particle equation in spherical polar coordinates. In (7), they are the corresponding solutions to this equation in prolate spheroidal coordinates. The lowest partial wave is of Σ_g^+ symmetry.

The function

$$p(\lambda) = \{1 - \exp[-\gamma(\lambda - 1)]\}^2$$

is a shielding function to make Cp continuous at $\lambda = 1$, the lowest value of this coordinate. $\{\phi_j \Psi_G\}_{j=1}^{N}$ are short-range correlation functions to describe the system when the positron is close to the H_2 target.

Ψ_G is an approximation to the ground-state of H_2 target wave function. It is included as a factor in Ψ_t to make possible the application of the method of models [14]. In this method, the target Hamiltonian $\hat{H}_{\text{target}}$ is replaced by a model Hamiltonian $\hat{H}_M$ where

$$\hat{H}_M = \hat{T}_{KE} + V_M,$$

$\hat{T}_{KE}$ is kinetic energy operator from $\hat{H}_{\text{target}}$ and V_M is a model potential such that

$$\hat{H}_M \Psi_G = E_M \Psi_G.$$

The energy E of the scattering state under consideration is taken to be of the form

$$E = E_M + \tfrac{1}{2}k^2,$$

where k is the wave number of the incident positron. (Here and elsewhere in this article the units used are Hartree atomic units.)

The method of models has the advantage that the only matrix elements we need evaluate involving $\hat{H}_M$ are those that involve $\hat{T}_{KE}$. In particular, we need not evaluate matrix elements of the Coulombic repulsion potential between the electrons. If particle 1 is the positron and the electrons are particles 2 and 3, this potential is $\dfrac{1}{r_{23}}$. This makes it unnecessary to calculate matrix elements involving $\dfrac{r_{12}r_{13}}{r_{23}}$ which arise when Hylleraas-type functions containing r_{12}, and by exchange r_{13}, are included in the short range correlation functions. This is a very important simplification. It is still necessary to evaluate matrix elements involving, for example, $\dfrac{r_{12}}{r_{13}}$ and $r_{12}r_{13}$, but this can be done much more easily.

Details of how the calculation was carried out are given in refs. [12] and [15]. The main points are as follows.

(1) The open channel functions S and C in equation (7) are separable functions of λ and μ. See, for example, ref. [3].

(2) In the integrals involving $\frac{r_{12}}{r_{13}}$ or $r_{12}r_{13}$, the electrons, particles 2 and 3, only occur in *one* factor. Also, as the H_2 nuclei were fixed at their equilibrium internuclear distance, $1.4a_0$, it was not necessary to include any of the μ coordinates in the exponent in the short-range correlation functions. As a consequence the double Neumann series expansions for $\frac{r_{12}}{r_{13}}$ and $r_{12}r_{13}$ terminate.

(3) The integrals in the expansions that involve electron coordinates can be evaluated analytically. The method used involved recursion formulae for the $Q_\tau^\nu(\lambda)$ functions. This can give rise to problems but the series terminate at sufficiently low τ values for the method to be acceptable.

(4) The final integration over λ_1 was carried out using Boys' boundary derivative reduction method [16,12].

(5) Test one-centre integrals were checked by analytical evaluation using the method of Roberts [17].

As the H_2 nuclei were fixed at their equilibrium value, no account was taken of vibrational-rotational excitation. As H_2 dissociation can be neglected for incident positron energies less than 8.8 eV, on account of the Frank–Condon principle [18], the first significant inelastic threshold is positronium formation at 8.63 eV. Thus our elastic scattering calculations were for energies less than this value.

A comparison between our results for the Kohn phase shift and the corresponding eigenphase sums obtained by Tennyson [19] using the R-matrix method is given in Table 9 in ref. [15]. The results for elastic cross sections obtained by more detailed calculations using both methods [13,20] are compared with experiment in Figure 1.

It can be seen that the Kohn calculation is in agreement with the results of Hoffman et al. [21] below 5 eV whereas the R-matrix calculation is in agreement with the results of Charlton et al. [22] in this range. The comparison between the results in Table 9 in ref. [15] indicates that the R-matrix results have not fully converged as even the results obtained by the most detailed application of this method are quite significantly below the results obtained using the Kohn method, *with the inclusion of Hylleraas-type basis functions.* This suggests quite strongly that the results of Hoffman et al. are more accurate than those of Charlton et al. for energies less than 5 eV.

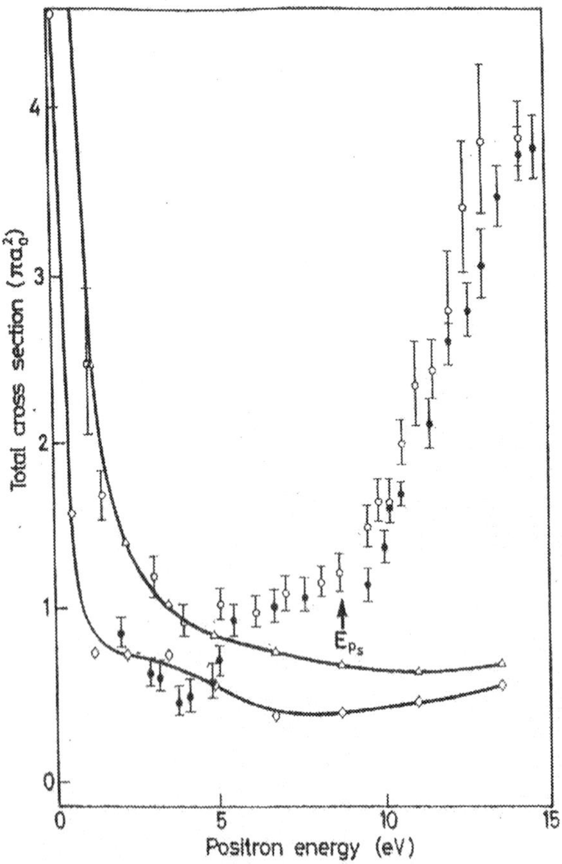

FIG. 1. Comparison of experimental and theoretical results: ○, Hoffman et al. [21]; •, Charlton et al. [22]; △, calculation [13] including Σ_g^+, Σ_u^+, Π_u and Π_g symmetries; ◇, Danby and Tennyson [20] including Σ_g^+, Σ_u^+, Π_u, Π_g, Δ_g and Δ_u symmetries.

IV. POSITRON ANNIHILATION

A phenomenon of interest to all of us involved in the CCP2 flagship project referred to in the introduction to these proceedings is *positron annihilation*. The probability of positron annihilation in e^+H_2 scattering is very small. The annihilation rate, λ_A, is given by

$$\lambda_A = \pi r_0^2 c N Z_{\rm eff}(k) \tag{8}$$

where

$$r_0 = \text{classical radius of the electron},$$
$$N = \text{density of } H_2 \text{ molecules in the vicinity of the positron},$$
$$k = \text{wave number of the incident positron}$$

and $Z_{\text{eff}}(k)$ = number of electrons available to the positron for annihilation.

$Z_{\text{eff}}(k)$ would be 2 for H_2, the number of electrons in H_2, if there were no interaction between the positron and H_2.

$$Z_{\text{eff}}(k) = \langle \Psi_t | \delta(\boldsymbol{r}_1 - \boldsymbol{r}_2) + \delta(\boldsymbol{r}_1 - \boldsymbol{r}_3) | \Psi_t \rangle, \tag{9}$$

where Ψ_t is normalised to one incoming positron per unit volume far away from the H_2.

Accurate evaluation of $Z_{\text{eff}}(k)$ requires Ψ_t to be accurate when the positron and an electron coincide. The exact wave function, Ψ_{ex}, at points of coincidence must satisfy the cusp condition [23,24]

$$\left(\frac{\partial \hat{\Psi}_{ex}}{\partial r_{1j}} \right)_{r_{1j}=0} = -\tfrac{1}{2} (\Psi_{ex})_{r_{1j}=0} \tag{10}$$

where $\hat{\Psi}_{ex}$ is the average value of Ψ_{ex} taken over a small sphere with r_{1j} = constant, with variables other than $\boldsymbol{r}_{1j}$ fixed. Hylleraas-type functions make it possible for Ψ_t to fit this condition quite accurately.

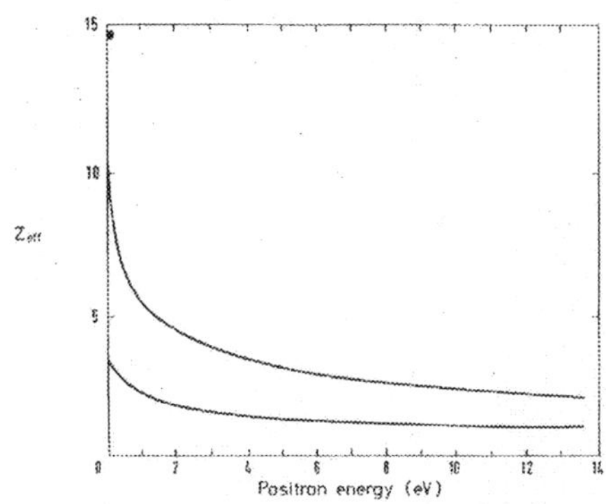

FIG. 2. Z_{eff} for positron-H_2 scattering calculated using the Kohn method and the lowest spheroidal wave [15]. • Experimental: McNutt et al. [25].

The results obtained for $Z_{\text{eff}}(k)$ using the Kohn variational method are shown in Figure 2. The lower curve was obtained using 64 short-range correlation functions containing one electron functions of σ or π symmetry. The upper curve was obtained by adding 8 Hylleraas-type functions to this basis set.

It can be seen that including Hylleraas-type functions increases dramatically the calculated value of $Z_{\text{eff}}(k)$ at very low energies but not sufficiently to agree with experiment. We consider that an important first step in trying to carry out calculations to account for the very high Z_{eff} value that has been observed, for example, in positron-acetylene scattering [26] is to carry out an accurate evaluation of $Z_{\text{eff}}(k)$ for e^+H_2 scattering at very low energies.

In what follows, I will outline how this can be combined with an extension of the calculation to include the positronium formation channel.

V. CALCULATIONS INVOLVING ANTIHYDROGEN

Over the past few years, together with others I have carried out calculations of the interaction of antihydrogen ($\bar{\text{H}}$) with H and He atoms. See, for example, refs. [27,28]. The first preparation of cold $\bar{\text{H}}$ at CERN in 2002 has given added impetus to this research [29,30].

Calculations on He-$\bar{\text{H}}$ involve the same three light particles, a positron and two electrons, as in the case of e^+H_2. We have carried out variational calculations of the Born–Oppenheimer potential for He-$\bar{\text{H}}$ using basis sets containing CI functions made up of products of one particle σ or π functions and also Hylleraas-type functions containing r_{12} and by exchange r_{13}.

In these calculations, it was necessary to evaluate integrals involving $\dfrac{r_{12}r_{13}}{r_{23}}$, in addition to those involving $\dfrac{r_{12}}{r_{13}}$ or $r_{12}r_{13}$.

The methods used in the e^+H_2 calculations were improved. More use was made of 1D numerical integration and more accurate methods were used for evaluating the Q_τ^ν functions [31]. Use was made of the inverse powers expansion

$$Q_\tau^\nu(\lambda) = (\lambda^2-1)^{\frac{\nu}{2}} \sum_{m=\tau+\nu+1}^{\infty} \frac{b_m(\tau,\nu)}{\lambda^m}, \quad |\lambda| > 1$$

where $b_m(\tau,\nu)$ is a coefficient. This expansion is slowly convergent for λ close to 1. However, recursion gives accurate values in this region.

I wrote a large program to evaluate the integrals involving $\dfrac{r_{12}r_{13}}{r_{22}}$. This required a triple Neumann expansion (4). No termination occurred and many terms contributed. However, it was possible to carry out the evaluations using only 2D numerical integration. To get accurate results, it was necessary to take τ in each expansion to be at least as high as $\tau = 8$. The accuracy of the results for some single-centre test integrals was checked using values obtained by Peter Van Reeth, using analytical methods he and Humberston developed to carry out the calculations described in ref. [11]. Martin Plummer is currently investigating how this program can be run on an HPC.

The results obtained at internuclear distance $R = a_0$ for the energy of the three light particles and the delta function expectation value

$$\langle \delta_{ee^+} \rangle = \langle \Psi_{BO} | \delta(\boldsymbol{r}_1 - \boldsymbol{r}_2) + \delta(\boldsymbol{r}_1 - \boldsymbol{r}_3) | \Psi_{BO} \rangle, \tag{11}$$

where Ψ_{BO} is the normalised, Born–Oppenheimer wave function, are given in Table 1. The maximum number of basis functions included was 164. They are compared with the results obtained by Strasburger and Chojnacki [32,33] using a basis set made up of 768 explicitly correlated Gaussian (ECG) functions

$$\chi_l = S \exp\left[-\sum_{i=1}^{3} \alpha_i^{(l)} (\boldsymbol{r}_i - \boldsymbol{R}_i^{(l)})^2 - \sum_{i>j}^{3} \beta_{ij}^{(l)} (\boldsymbol{r}_i - \boldsymbol{r}_j)^2 \right],$$

where S is the symmetriser for the electrons.

Type and number of basis functions					
σ	$\pi_1\pi_2$ and $\pi_1\pi_3$	$\pi_2\pi_3$	r_{12} and r_{13}	Energy	$\langle \delta_{ee^+} \rangle$
84	28	28	0	−1.6125954	0.02400
84	28	28	1	−1.6212778	0.03514
84	28	28	15	−1.6295878	0.04614
84	28	28	24	−1.6307149	0.04763
768 ECG functions [32,33]				−1.6364739	0.05521

Table 1 Energy and delta function expectation values for $\mathrm{He} - \bar{\mathrm{H}}$ at $R = a_0$. Units are atomic units.

It can be seen that the energy value obtained by Strasburger and Chojnacki is lower and therefore more accurate than our value. Also their value for $\langle \delta_{ee^+} \rangle$ is 16% larger than our value obtained with 24 Hylleraas-type functions. Their value for $\langle \delta_{ee^+} \rangle$ for $\mathrm{He} - \bar{\mathrm{H}}$ at $R = 0$, where it corresponds to positronium hydride (PsH), is in good agreement with the value obtained by Ho [34] for PsH using 396 basis functions, some of which were Hylleraas-type functions.

Thus we conclude that the ECG basis gives accurate results for $\langle \delta_{ee^+} \rangle$, more accurate than we have obtained so far using Hylleraas-type basis functions. This is in spite of the fact that it is not possible to satisfy the cusp condition (10) using ECG basis functions.

We can improve the accuracy of our calculation by including more basis functions and also by including Hylleraas-type functions containing the interelectron distance, r_{23}. The results obtained for $\mathrm{H}-\bar{\mathrm{H}}$ [27] indicate that we are unlikely to obtain results that are quite as accurate as those obtained using an ECG basis set [32]. However, it was reasonably straightforward to extend our calculation to calculate T-matrix elements for the $\mathrm{He}^+\bar{\mathrm{p}}+\mathrm{Ps}$ rearrangement channel [28], using 5-dimensional numerical integration and a distorted wave approximation.

VI. ADAPTATION TO e^+ H_2 SCATTERING

As part of the flagship project, we intend to adapt the $\mathrm{He}-\bar{\mathrm{H}}$ calculation to make possible a Kohn calculation of $e^+\mathrm{H}_2$ scattering including the positronium formation channel. Replacement of the nuclei in $\mathrm{He}-\bar{\mathrm{H}}$ by protons will make possible calculation of matrix elements involving only short-range correlation functions without the use of the method of models [14]. Matrix elements involving open channel functions will have to be calculated, but these will not involve $\frac{r_{12}r_{13}}{r_{23}}$. The T-matrix element calculations involving $\mathrm{He}^+\bar{\mathrm{p}}+\mathrm{Ps}$ can be adapted to make possible the calculation of matrix elements between the $e^++\mathrm{H}_2$ entrance channel and the positronium formation channel, $\mathrm{H}_2^+ + \mathrm{Ps}$.

VII. CONCLUSION

I have described how the method devised by Hylleraas for carrying out an accurate calculation of the ground-state energy of helium [1] and later extended to molecules containing up to 4 electrons [2,5,6,7] can be used to carry out Kohn calculations of elastic $e^+\mathrm{H}_2$ scattering [13]. These are similar to calculations that have been carried out for $e^+\mathrm{He}$ [11,12].

These calculations give accurate results for the total cross section up to 5 eV and a much improved value of Z_{eff} over earlier CI calculations, but still significantly below the experimental value. The results for the cross section are considerably more accurate than those obtained using the R-matrix method, which used basis sets made up of CI functions [20,15].

I have also described how the method has been applied to the interaction between helium and antihydrogen. We have carried out calculations of the Born–Oppenheimer energy and the $\langle\delta_{ee^+}\rangle$ expectation value (11) for $\mathrm{He}-\bar{\mathrm{H}}$ by this method. These results have not fully converged. In both cases considered they are less accurate than the results obtained by Strasburger and Chojnacki using a larger, explicitly correlated Gaussian basis set [32]. In the case of $\langle\delta_{ee^+}\rangle$, this was despite the fact that an ECG basis cannot fit the cusp condition (10).

Some improvement in the accuracy of our results will be possible by extending the basis set, but it is doubtful if they can be made quite as accurate as the results obtained by Strasburger and Chojnacki. However, we have successfully applied the method to the calculation of T-matrix elements for the rearrangement of $\mathrm{He} + \bar{\mathrm{H}}$ into $\mathrm{He}^{+}\bar{\mathrm{p}} + \mathrm{Ps}$ [28].

The $\mathrm{He} - \bar{\mathrm{H}}$ BO energy calculation will be adapted to make possible improved calculations of $e^{+}\mathrm{H}_2$ scattering, with inclusion of the positronium formation channel. This calculation will be used to evaluate Z_{eff}. It is hoped that a value in agreement with experiment can be obtained.

The insights on how to take into account the positron-electron interaction accurately that have been gained from this calculation, together with the most up-to-date methods of treating electron-electron correlation in quantum chemical calculations described by many speakers at this Workshop, will be very useful for the flagship project. As indicated in the introduction to these proceedings and by Tennyson and Franz [35], they will be used to try to devise an R-matrix method of calculating accurately the very high Z_{eff} values that have been observed in positron scattering by molecules such as acetylene [26]. These very high values are considered to be due to a resonant process involving the vibrational motion of the nuclei of the molecule [36].

Acknowledgements

I wish to thank John Humberston, Peter Van Reeth and Jonathan Tennyson for valuable discussions over many years and SERC and EPSRC for support for this research.

---

[1] E.A. Hylleraas, *Z. Physik* **54**, 347 (1929); *Rev. Mod. Phys.* **35**, 421 (1963).
[2] H.M. James and A.S. Coolidge, *J. Chem. Phys.* **1**, 825 (1933).
[3] C. Flammer, *Spheroidal Wave Functions* (Stanford University Press, Stanford, CA, 1957).
[4] F.E. Neumann, *J. Reine Angew. Math. (Crelle)* **37**, 21 (1848); Vorlesungen über die Theorie des Potentials und der Kugel funktionen (Teubner, Leipzig, 1878) Chap. 13.
[5] W. Kołos and C.C.J. Roothaan, *Rev. Mod. Phys.* **32**, 219 (1960).
[6] W. Kołos, *Int. J. Quant. Chem.* **2**, 471 (1968).
[7] D.C. Clary, *Molec. Phys.* **34**, 793 (1977).
[8] E.V. Rothstein, *Phys. Rev. A* **3**, 1581 (1971).
[9] H.S.W. Massey and R.O. Ridley, *Proc. Phys. Soc. (London) A* **69**, 659 (1956).
[10] H.S.W. Massey and A.H.A. Moussa, *Proc. Phys. Soc. (London) A* **71**, 38 (1958).

[11] J.W. Humberston and P. Van Reeth, these proceedings.
[12] E.A.G. Armour and J.W. Humberston, *Physics Reports* **204**, 165 (1991).
[13] E.A.G. Armour, D.J. Baker and M. Plummer, *J. Phys. B* **23**, 3057 (1990).
[14] R.J. Drachman, *J. Phys. B* **5**, L30 (1972).
[15] E.A.G. Armour, *Physics Reports* **169**, 1 (1988).
[16] N.C. Handy and S.F. Boys, *Theor. Chim. Acta* **31**, 195 (1973).
[17] P.J. Roberts, *J. Chem. Phys.* **43**, 3547 (1965).
[18] H.S.W. Massey, *Electronic and ionic impact phenomena* (OUP, Oxford, 1969) p.877.
[19] J. Tennyson, *J. Phys. B* **19**, 4255 (1986).
[20] G. Danby and J. Tennyson, *J. Phys. B* **23**, 1005, 2471 (1990).
[21] K.R. Hoffman, M.S. Dababneh, Y.-F. Hsieh, W.E. Kauppila, V. Pol and T.S. Stein, *Phys. Rev. A* **25**, 1393 (1982).
[22] M. Charlton, T.C. Griffiths, G.R. Heyland and G.L. Wright, *J. Phys. B* **16**, 323 (1983).
[23] T. Kato, *Commun. Pure Appl. Math.* **10**, 151 (1957).
[24] R.T. Pack and W. Byers Brown, *J. Chem. Phys.* **45**, 556 (1966).
[25] J.D. McNutt, S.C. Sharma and R.D. Brisbon, *Phys. Rev. A* **20**, 347 (1979).
[26] L.D. Barnes, S.G. Gilbert and C.M. Surko, *Phys. Rev. A* **67**, 032706 (2003).
[27] E.A.G. Armour and C.W. Chamberlain, *J. Phys. B* **35**, L489 (2002).
[28] E.A.G. Armour, S. Jonsell, Y. Liu and A. Todd, *Nucl. Instr. and Meth. B*, in press (2006).
[29] M. Amoretti et al., *Nature* **419**, 456 (2002).
[30] G. Gabrielse et al., *Phys. Rev. Lett.* **89**, 213401, 233401 (2002).
[31] C.P. Franklin, PhD thesis, University of Nottingham (1994).
[32] K. Strasburger and H. Chojnacki, *Phys. Rev. Lett.* **88**, 163201 (2002).
[33] K. Strasburger, *J. Phys. B* **37**, 2211 (2004).
[34] Y.K. Ho, *Phys. Rev. A* **34**, 609 (1986).
[35] J. Tennyson and J. Franz, these proceedings.
[36] G.F. Gribakin, *Phys. Rev. A* **61**, 022720 (2000).

E. A. G. Armour, J. Franz and J. Tennyson (eds.)
Explicitly Correlated Wavefunctions
© 2006, CCP2, Daresbury

Low-energy electron and positron collisions with the R-matrix method

Jonathan Tennyson and Jan Franz

Department of Physics and Astronomy, University College London, London WC1E 6BT, UK

I. INTRODUCTION

The R-matrix method is a well established computational procedure for treating electron collisions with atoms and molecules [1]. The basis of the method is the division of space into two regions: an inner region, defined by a sphere typically of radius 10 to 15 a_0, and an outer region. The inner region must contain the entire electron density of the target system which means that it is only in this region that complicated correlation and exchange processes need to be treated in detail. The R-matrix provides the link between the two regions.

For moleular targets, the calculations within the inner region reduce to a modified electronic structure calculation and standard quantum chemistry codes have been adapted for the purpose. In particular the UK molecular R-matrix codes [2] uses adapted versions [3,4] of the Alchemy and Sweden-Molecule codes for diatiomic and polyatomic targets respectively. All calculations discussed here used this code, or updated versions of it.

In the inner region the wavefunction for an $N+1$ electron scattering problem is represented by the close-coupling expansion:

$$\Psi_k = \mathcal{A} \sum_{ij} a_{ijk} \phi_j^N \eta_{ij} + \sum_i b_{ik} \chi_i^{N+1}, \qquad (1)$$

where ϕ_j^N is the wavefunction of jth state of the N-electron target, usually represented by a complete active space configuration interaction (CAS-CI) expansion of up to a few thousand terms expressed in natural orbitals. η_{ij} is one-electron function representing the scattering electron which is antisymmetrised to the target by operator $\mathcal{A}$. The terms in the second summation are expressed entirely in terms of target orbitals and are necessary to relax the orthogonalisation condition between the target and continuum orbitals; other effects such as polarisation can be included in this term. The coefficitons a_{ijk} and b_{ik} are variational and are determined by diagonalising the secular matrix. Changes to this form needed for low-energy positron collisions are considered below.

II. ELECTRON COLLISION CALCULATIONS

The UK molecular R-matrix code has demonstrated that it can obtain reliable results for a variety of processes involving low energy scattering. In particular studies on elastic scattering and rotational excitation of water [5], electronic excitation of carbon monoxide [6] and vibrationally-resolved resonance effects in electronic excitation of molecular hydrogen [7] have all given results competitive with and, in at least one case [5], probably better than experiment. Experience has shown for these studies that the key to getting good results is obtaining a reliable and flexible representation of the target. That is to say, within limits imposed by the standard model used [8] and computer power, the low-energy electron-molecule scattering problem is one that is numerically solved.

Recent developments in electron scattering studies have focussed on extending the calculations to larger systems and higher energies. Studies have been performed on the electron rich CF_3 radical [9] and the small organic ring system tetrahydrofuran (THF) C_4H_8O [10]. These studies show that the computer resources required for treating such systems at a fairly sophisticated level are not excessive but there remains issues to do with constructing reliable target models. In particular the results of both studies proved to be sensitive to the choice of target wavefunctions. One issue here is that all target states used in the close-coupling expansion, eq. (1), have to be represented within a common basis set. For larger systems detailed electronic structure calculations on prospective targets, particularly calculations which consider excited states with different spins to the ground state, are often not available. It is likely that performing such calculations at some high level of theory will become a routine part of future electron scattering studies.

For electron scattering the collisions energies can usefully be divided into three distinct regions. The low-energy region where close-coupled expansions like that of eq. (1) are appropriate, the high-energy regime where collisions are essentially impulsive and methods based on perturbation theory give good results. The intermediate energy region spans the ionisation energy of the target. At ionisation the number of open (energetically accessible) states that need to be included in expansion (1) becomes infinite and near the threshold for ionisation the electron is not moving fast enough for post-collision electron-target interactions to be neglected.

There has recently been significant progress addressing the intermediate energy problem for atomic targets. At least four *ab initio* procedures have been developed for treating these problems with considerable success. However most of these procedures have only been implemented for simplified one-electron or quasi one-electron targets. The most flexible, if not the most elegant procedure, is the R-matrix with pseudo-states (RMPS) method [11]. Recently Gorfinkiel and Tennyson [12,13] have developed a related molecular R-matrix with pseudo-states (MRMPS) method which is the first *ab initio* procedure for treating near-threshold electron impact ionisation of molecules.

In the MRMPS method, the close-coupling expansion of eq. (1) is augmented by a (hopefully near-complete) series of extra (pseudo) states. These extra states are not meant to correspond to physical states of the system but to represent a discretised version of the continuum (and any missing excited states) within the space of the R-matrix sphere. Gorfinkel and Tennyson used even tempered Gaussian type orbitals placed on the molecular centre-of-mass to represent these pseudo-states. Even tempered functions have the advantage that they can, at least in principle, be extended to give a complete set and several different series can be defined which is important for both demonstrating the stability of the calculation and dealing with the many artificial (pseudo-)resonances which are often a consequence of the procedure. So far MRMPS calculations have been performed successfully for electron collisions with H_3^+ [12,13] and H_2 [13]. The challenge is now to apply this method to problems involving many electron targets.

III. POSITRON COLLISION CALCULATION

Superficially the inner region R-matrix wavefunction of eq. (1) can be adapted to problems involving positron collisions by removing the anti-symmetriser, changing the sign of appropriate Coulomb interactions and by distinguishing between the target electrons and scattering positron. While this is formally correct, it pays scant attention to the extra difficulties encountered in calculations which mix positrons and electrons. The particular problem here is that the Coulomb *attraction* between these particules leads to the need for very careful treatment of electron-positron correlation problem. This is particularly true for low-energy scattering where the slow-moving positron can easily react to the motions of the target electrons, and in the region of the positron formation threshold where the positron can rip an electron from the target to form the hydrogen-like positronium atom.

Two decades ago the then diatomics-only molecular R-matrix code was adapted to treat low-energy positron-molecule collision [14]. This code was applied to collisions between positrons and molecular hydrogen [15] and molecular nitrogen [16]. Even though these studies led to the first-ever prediction of a positron-molecule bound state [17], they were of limited usefulness because the of the low level treatment of the positron-electron correlation problem.

Since these studies were performed the R-matrix codes have not only been extended to treat polyatomic targets but also been re-written to allow for the use of significantly more complicated representations of the molecular target and hence associated scattering models [18]. We have undertaken a re-implentation of positron collision effects within the modern UK molecular R-matrix codes, some first results of which are given below. So far our implementation remains within the close-coupling paradigm of eq. (1). We recognise that this form of the wavefunction is inadequate for important aspects of positron scattering but it represents an important staging post on the road to a more sophisticated and more reliable *ab initio* treatment of this challenging problem.

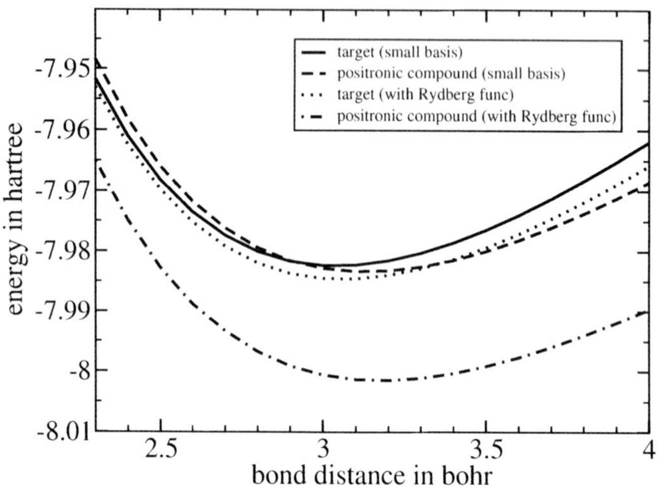

FIG. 1. Positron binding to LiH as a function of bond length

The positronic bound state of molecular LiH provides an excellent test of our newly implemented method. A series of studies on this system using fairly standard [19–22] and quantum Monte Carlo [23,24] "electronic" structure treatments are available. Figure 1 presents our calculated potential curves for the LiH target and the associated lowest state of e^+LiH. These calculations are purely bound state studies ie without an R-matrix sphere. The upper curves uses merely a standard basis (DZP (Dunning): (9s4p1d/4s2p1d) for Li and (5s2p/3s2p) for H) to represent the target and compound system. Within this model the positron is barely bound. The lower curves augmented this basis with diffuse functions (5s functions with exponents: 0.0100, 0.0050, 0.0025, 0.0010, and 0.0005; 4 p functions with exponents 0.0090, 0.0060, 0.0030, and 0.0009; 1 d function with exponent 0.056). As expected for a variational procedure, this lowers the target energy. However the energy of the compound system is lowered significanly more leading to the well bound system predicted by the studies quoted above. At the level of correlation used in our present codes, which takes into all single excitations only, this calculation cannot be expected to be fully quantitative but it certainly reproduces the qualitative features of the more sophisticated studies cited above.

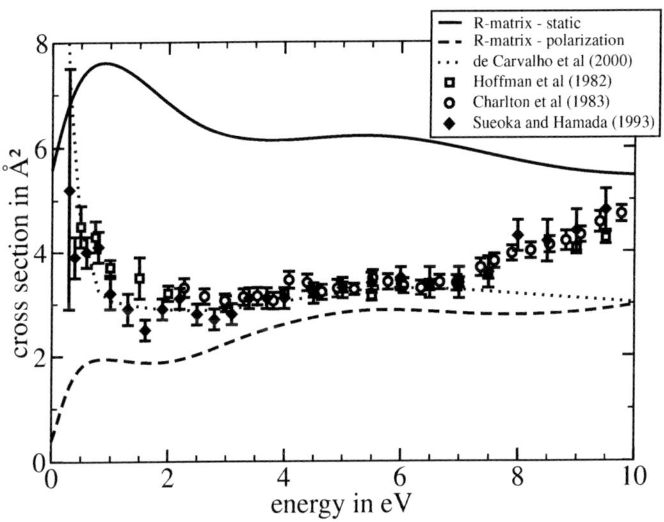

FIG. 2. Cross-section for positron-N_2 scattering

Figure 2 gives results for a number of studies on elastic scattering of positrons by molecular nitrogen. This collision problem is of particular importance since N_2 is the gas usually employed in positron traps to thermalise the positrons. As is well-know from earlier studies [16], the static approximation leads to cross section which are systematically too high below the positronium formation threshold. This approximation overestimates the repulsion between the target and the incoming positron since the target is frozen and unable to polarise. Inclusion of this polarisation is altogther trickier and, unlike the static approximation, shows a strong model dependence. Figure 2 shows that using the model based on eq. (1) performs reasonable well at the higher energies considered, which is of comparable quality with earlier R-matrix studies [16], but is unable to reproduce the correct behaviour at the lowest energies. This is due to the incomplete representation of polarisation effects.

IV. CONCLUSIONS

As is well-known the inclusion of polarisation (or correlation) effects in low-energy positron collision problems is significantly more demanding than in the corresponding electron collisions. The treatment of correlation effects in the calculations of the electronic structure of small molecules has made huge recent strides by employing methods which explicitly treat the electron-electron coordinate [27]. Such methods have been employed very successfully by Armour and co-workers for positron-H_2 collision problems [25]. It would seem that for significant progress to be made, this approach should be generalised, perhaps by adapting the formalism of Gdanitz [26]. Such a development should prove particularly important for treating the interesting and not well understood positron annihilation problem.

Acknowledgement

The UK molecular R-matrix code has been supported over many years by UK Collaborative Computational Project 2 (CCP2). The present project is funded by EPSRC as the CCP2 flagship project. We thank chair of CCP2, Edward Armour, both for encouraging us to undertake this work and for many helpful discussion.

---

[1] P G Burke and K A Berrington, editors. *Atomic and Molecular Processes, an R-matrix Approach.* Institute of Physics Publishing, Bristol, 1993.

[2] L. A. Morgan, J. Tennyson, and C. J. Gillan. *Computer Phys. Comms.*, 114:120–128, 1998.

[3] C. J. Noble. *The ALCHEMY Linear Molecule Integral Generator.* DL/SCI/TM33T Daresbury Laboratory Technical Memorandum, 1982.

[4] L. A. Morgan, C. J. Gillan, J. Tennyson, and X. Chen. *J. Phys. B: At. Mol. Opt. Phys.*, 30:4087–4096, 1997.

[5] A. Faure, J. D. Gorfinkiel, and J. Tennyson. *J. Phys. B: At. Mol. Opt. Phys.*, 37:801–807, 2004.

[6] L. A. Morgan and J. Tennyson. *J. Phys. B: At. Mol. Opt. Phys.*, 26:2429–2441, 1993.

[7] D. T. Stibbe and J. Tennyson. *Phys. Rev. Lett.*, 79:4116–4119, 1997.

[8] J. Tennyson. *J. Phys. B: At. Mol. Opt. Phys.*, 29:6185–6201, 1996.

[9] I. Rozum, N. J. Mason, and J. Tennyson. *New J. Phys.*, 5:1–155, 2003.

[10] D Bouchiha, J D Gorfinkiel, L G Caron, and L Sanche. *J. Phys. B: At. Mol. Opt. Phys.*, 39(4):975, 2006.

[11] K. Bartschat, E. T. Hudson, M. P. Scott, P. G. Burke, and V. M. Burke. *J. Phys. B: At. Mol. Opt. Phys.*, 29:115, 1996.

[12] J. D. Gorfinkiel and J. Tennyson. Electron-H_3^+ collisions at intermediate energies. *J. Phys. B: At. Mol. Opt. Phys.*, 37:L343–L350, 2004.

[13] J. D. Gorfinkiel and J. Tennyson. *J. Phys. B: At. Mol. Opt. Phys.*, 38:1607–1622, 2005.

[14] J. Tennyson. *J. Phys. B: At. Mol. Phys.*, 19:4255–4263, 1986.

[15] G. Danby and J. Tennyson. *J. Phys. B: At. Mol. Opt. Phys.*, 23:1005–1016, 1990. erratum 23, 2471 (1990).

[16] G. Danby and J. Tennyson. *J. Phys. B: At. Mol. Opt. Phys.*, 24:3517–3529, 1991.

[17] G. Danby and J. Tennyson. *Phys. Rev. Letts.*, 61:2737–2739, 1988.

[18] J. Tennyson. A new algorithm for hamiltonian matrix construction in electron-molecule collision calculations. *J. Phys. B: At. Mol. Opt. Phys.*, 29:1817–1828, 1996.

[19] K Strasburger. *J. Chem. Phys.*, 111:10555, 1999.

[20] K Strasburger. *J. Chem. Phys.*, 114:615, 2001.

[21] R J Buenker, H-P Liebermann, V Melnikov, M Tachikawa, L Pichl, and M Kimura. *J. Phys. Chem. A*, 109:5956, 2005.

[22] F A Giantruco, J Franz, R J Buenker, H-P Liebermann, L Pichl, J-M Rost, M Tachikawa, and M Kimura. *Phys. Rev. A*, 73:022705, 2006.

[23] M Mella, G Morosi, D Bressanini, and S Elli. *J. Chem. Phys.*, 113:6154, 2000.

[24] M Mella, S Chiesa, and G Morosi. *J. Chem. Phys.*, 116:2852, 2002.

[25] E A G Armour and J W Humberston. *Phys. Rep.*, 204:165, 1991.

[26] R J Gdanitz. *Chem. Phys. Lett.*, 210:253, 1993.

[27] W Klopper. *R12 methods, Gaussian geminals* in: Modern Methods and Algorithms of Quantum Chemistry, Proceedings Winterschool 21-25 February 2000, Second Edition, NIC Series, Vol. 3, J. Grotendorst (Ed.), Forschungszentrum Jülich, Germany, 2000, pp. 181-229

E. A. G. Armour, J. Franz and J. Tennyson (eds.)
Explicitly Correlated Wavefunctions
© 2006, CCP2, Daresbury

Variational calculations on positronic bound states

Krzysztof Strasburger and Małgorzata Wołcyrz

Institute of Physical and Theoretical Chemistry, Wrocław University of Technology, Wybrzeże Wyspiańskiego 27, 50-370 Wrocław, Poland
strasbur@chkw386.ch.pwr.wroc.pl

In the last decade of 20'th century a breakthrough in the studies of positronic matter has happened. This recent progress is mainly due to two computational methods: Quantum Monte Carlo simulations and variational calculations with the basis of explicitly correlated Gaussian (ECG) functions. The latter method provides very accurate energies and good annihilation rates of small positronic atoms and molecules (practically, containing up to 4 electrons). Studies of larger positronic atoms with one or two valence electrons were also performed, with atomic cores excluded from the explicitly correlated part of the wave function, while the core-valence and core-positron correlations were represented by polarization potentials. Finally, accurate adiabatic interaction energies for interactions of antihydrogen atoms with atomic and molecular hydrogen, and with helium in singlet and triplet states should be mentioned.

Studies of positronic matter with ECG functions are carried out, among others, in our laboratory. The first part of the present text is devoted to this kind of wave function and respective computational techniques. On the other hand, these calculations are very time consuming. There is still no efficient method, which would make it possible to perform routine calculations on larger positronic molecules, giving reasonable annihilation rates. An approach, which is based on adiabatic separation of electronic and positronic degrees of freedom and could fulfill this requirement for a wide class of positronic molecules, will be introduced in the second part.

I. VARIATIONAL CALCULATIONS WITH EXPLICITLY CORRELATED GAUSSIAN FUNCTIONS

The wave function should possess a maximum at the coalescence point for an electron-positron pair, which has zero relative angular momentum. Accordingly to the cusp condition [1],

$$\lim_{r_{ep}\to 0} \frac{\partial \psi}{\partial r_{ep}} = -\frac{1}{2} \tag{1}$$

there is a discontinuity of first derivative of the wave function at this point. Such a maximum would be best reproduced by a factor, having the form of $e^{-\alpha r_{ep}}$, which would resemble the wave function of positronium. Unfortunately, wave functions depending exponentially on interparticle distance, lead to complicated integrals, if more than 3 particles are involved. The cusp can be also well reproduced by simple linear factor, which appears as the second term in power expansion of exponential function. The efficiency of such a correlation factor is known since the early work of Hylleraas [2]. Trial wave functions, depending on powers of interelectronic distance, provided many valuable results for few electron atoms and two-electron molecules. However, ansatz with linear dependence on electron-positron distance has failed to predict the existence of a bound states in positronic lithium [3] and metastable positronic helium [3,4], both having the structure of a loosely bound positronium, polarized in the electric field of charged atomic core. It appears, that the ability to fulfill the cusp condition alone by the basis functions, does not guarantee fast convergence of the expansion towards the solution of the Schrödinger equation. It has been shown, that power dependence on the electron-positron distance leads to spurious nodal hypersurfaces and maxima of the density [5]. This deficiency is shared between CI and Hylleraas-type functions.

Therefore, it is important that correlation factors, occuring in basis functions used for positronic atoms and molecules, not only possess a maximum at the coalescence point, but also decrease to zero, as the interparticle distance increases. Such an ansatz, in the form of a linear combination of explicitly correlated Gaussian functions, has been proposed by Boys [6] and Singer [7] (in different, but mathematically equivalent forms). We are using the Boys form,

$$\Psi = \sum_{I=1}^{K} C_I \hat{A} e^{-\sum_{i=1}^{N+\overline{N}} \alpha_i^{(I)} (\mathbf{r}_i - \mathbf{R}_i^{(I)})^2 - \sum_{i>j}^{N+\overline{N}} \beta_{ij}^{(I)} (\mathbf{r}_i - \mathbf{r}_j)^2} \Theta_I, \tag{2}$$

where $\hat{A}$ is the appropriate antisymmetrizer and Θ denote spin functions. This kind of wave function was for many years regarded as inferior to Hylleraas-type functions, as it is unable to describe the cusp (only even powers of r_{ij} appear in the power expansion of correlation factors). The great advantage of ECG basis is that all integrals, needed for calculation of nonrelativistic energy, can be expressed by analytical formulas and the most complicated of them involve the error function. This simplicity opens the door to efficient optimization of nonlinear parameters α, β and $\mathbf{R}$. It has been demonstrated, that such an optimization is crucial for obtaining energies of high accuracy, which is comparable or even better than that given by Hylleraas (or James-Coolidge and Kołos-Wolniewicz, for molecules) functions, both for normal and positronic systems (see [8,9] for a review). Moreover, the ECG function is the one explictly correlated ansatz, which can be applied for molecules of any geometry. The numbers of electrons (N) and positrons ($\overline{N}$) are limited only by available computational resources.

$PsLi^+$		e^+LiH, R=3.365 bohr	
K	E (hartree)	K	E (hartree)
1504	-7.5324063	1024	-8.107563
2000	-7.5034077[1]	2048	-8.107865

[1]not converged yet

Table 1: Variational energies of $PsLi^+$ and e^+LiH

For the optimization of nonlinear parameters, a method described in reference [10] is used. Parameters of a single, chosen basis function are optimized simultaneously with Powell's method of conjugate directions. Such a limitation has been introduced in order to avoid recalculation of full H and S matrices in each step. Only a single row and column of these matrices have to be updated. After the energy is minimized with respect to parameters of the actually optimized function, the procedure moves onto next basis function. The whole process is stopped, when the energy is lowered by less than a given threshold in a big cycle, which passes through all basis functions. In table 1 results for positronic lithium ($PsLi^+$) atom and positronic lithium hydride (e^+LiH) molecule at its equilibrium geometry are presented. The energies are lowest reported to date for these systems. While the energy of $PsLi^+$ has converged to the accuracy of about 1 μhartree, the error for e^+LiH is by two orders of magnitude larger at similar expansion length.

Calculations with fully correlated wave functions are unlikely to be carried out for large systems, as their computational cost increases very quickly with the number of active particles. The most time consuming part of the algorithm is usually the calculation of two-particle integrals. For a system containing N electrons and $\overline{N}$ positrons, described with K-term ECG function, it scales proportionally to $K^2 N! \overline{N}! (N + \overline{N})^4$. Increasing complexity of matrix elements is not the one reason for longer computation times, as for a larger system, more basis functions are required for the results of the same accuracy, than for a smaller one. Furthermore, each basis function depends on a larger number of parameters, so that the energy has to be computed more times in the process of optimization.

II. ADIABATIC SEPARATION OF ELECTRONIC AND POSITRONIC DEGREES OF FREEDOM

Yet another approximate method of solving the leptonic Schrödinger equation for positronic atoms and molecules (containing only 1 positron) is currently being developed in our laboratory. For given, fixed nuclear coordinates, the leptonic wave function is assumed to be a product of positronic and electronic functions,

$$\Psi_{lep}(\mathbf{r_p}, \mathbf{r_e}) = \Psi_p(\mathbf{r_p})\Psi_e(\mathbf{r_p}, \mathbf{r_e}). \tag{3}$$

This approach could be called as second stage of adiabatic approximation (the first stage would be the obvious separation of nuclear and leptonic motions). Although the idea of not treating the positron on equal footing with electrons seems to be wrong, as the masses of particles are equal, it may appear to be not as crazy as it looks like. Treating the positron as a pseudonucleus and centering basis functions on it is, in some sense, equivalent to using a correlated ansatz. The Coulomb hill of the electron density on the positron, which is hardly available to methods based on configurational expansion, occurs in a natural way, giving a hope for good annihilation rates. The adiabatic approximation works well, if the separated subsystems move in different time scales. This assumption may be fulfilled quite well for a wide class of atoms and molecules, which do bind the positron weakly. The positron, in such bound states, occupies outer areas of the atom or molecule and moves slower than the electrons.

The first problem to be addressed is an optimal separation of the leptonic hamiltonian to electronic and positronic parts,

$$\hat{H}_{lep} = \hat{H}_e + \hat{H}_p, \tag{4}$$

for which fulfilling of the electronic Schrödinger equation by the electronic wave function is required,

$$\hat{H}_e\Psi_e(\mathbf{r_p}, \mathbf{r_e}) = E_e(\mathbf{r_p})\Psi_e(\mathbf{r_p}, \mathbf{r_e}). \tag{5}$$

Assuming that $\hat{H}_p = \hat{T}_p$ (i.e. the operator of kinetic energy of the positron), similarly to the adiabatic separation of nuclear motions, which is discussed in handbooks of quantum chemistry, is not the one possibility. For example, Mohallem proposed a method, which includes $\hat{T}_p$ into $\hat{H}_e$ [11]. Boroński and Stachowiak developed another approach, which treats the positron as a light nucleus having the charge of $\frac{1}{2}$, aimed at description of the positron in metal lattices [12]. There is nothing fundamental against exploitation of that idea for atoms and molecules, and its obvious generalization leads to treating the effective charge of the positron, q_p, as a variational parameter of the method. $\hat{H}_e$ becomes then

$$\hat{H}_e = -\frac{1}{2}\sum_{i=1}^{N_e}\nabla_i^2 - \sum_{\alpha=1}^{N_{nuc}}\sum_{i=1}^{N_e}\frac{Z_\alpha}{r_{\alpha i}} + \sum_{i>j}^{N_e}\frac{1}{r_{ij}} + \sum_{\alpha=1}^{N_{nuc}}\frac{Z_\alpha q_p}{r_{\alpha p}} - \sum_{i=1}^{N_e}\frac{q_p}{r_{ip}}$$

and the remaining part constitutes the $\hat{H}_p$ operator

$$\hat{H}_p = -\frac{1}{2}\nabla_p^2 + \sum_{\alpha=1}^{N_{nuc}} \frac{Z_\alpha(1-q_p)}{r_{\alpha p}} - \sum_{i=1}^{N_e} \frac{1-q_p}{r_{ip}}$$

Insertion of wave function (3) in the Schrödinger equation and integration over electronic coordinates leads to the equation for postronic function

$$\left[-\frac{1}{2}\nabla_p^2 + E_e(\mathbf{r_p}) + V_{es}(\mathbf{r_p}) + E_{ad}(\mathbf{r_p})\right]\Psi_p(\mathbf{r_p}) = E_{lep}\Psi_p(\mathbf{r_p})$$

The electronic energy, E_e, follows from equation (5) and has no physical meaning. The electrostatic correction, V_{es}, contains the electrostatic interaction of *remaining* positronic charge, $1-q_p$, with the nuclei and electrons.

$$V_{es}(\mathbf{r_p}) = \sum_{\alpha=1}^{N_{nuc}} \frac{Z_\alpha(1-q_p)}{r_{\alpha p}} - \left\langle \Psi_e(\mathbf{r_p}, \mathbf{r_e}) \left| \sum_{i=1}^{N_e} \frac{1-q_p}{r_{ip}} \right| \Psi_e(\mathbf{r_p}, \mathbf{r_e}) \right\rangle$$

E_{ad} is the adiabatic correction

$$E_{ad} = \left\langle \Psi_e(\mathbf{r_p}, \mathbf{r_e}) \left| -\frac{1}{2}\nabla_p^2 \right| \Psi_e(\mathbf{r_p}, \mathbf{r_e}) \right\rangle$$

Small positron mass implies the inadequacy of the Born-Oppenheimer approximation, so that E_{ad} has to be evaluated. It is the one serious obstacle to be overcome, before the method discussed here could be routinely applied. As long as the electronic energy is obtained with a variational method and the adiabatic correction is calculated accurately, the leptonic energy also remains variationally bound.

Let us conclude with the remark, that any approximation should be first tested on small systems, for which accurate, benchmarks results from fully correlated calculations are available. Respective tests for model one- and two-electron atoms are currently carried out in our laboratory.

---

[1] T. Kato, Commun. Pure Appl. Math. 10, 151 (1957)
[2] E. Hylleraas, Z. Physik 54, 347 (1929)
[3] D. C. Clary, J. Phys. B: At. Mol. Phys. 9, 3115 (1976)
[4] R. J. Drachman, Y. K. Ho, S. K. Houston, J. Phys. B: Atom. Molec. Phys. 9, L199 (1974)

[5] K. Strasburger, Int. J. Quantum Chem. 79, 243 (2000)
[6] S. F. Boys, Proc. R. Soc. London Ser. A 258, 402 (1960)
[7] K. Singer, Proc. R. Soc. London Ser. A 258, 412 (1960)
[8] J. Rychlewski, J. Komasa, in "Explicitly Correlated Wave Functions in Chemistry and Physics", Kluwer Academic Publishers 2003, pp. 91-137; K. Strasburger, H. Chojnacki, *ibid.*,pp. 439-458
[9] J. Mitroy, M. W. J. Bromley, G. G. Ryzhikh, J. Phys. B 35, R81 (2002)
[10] J. Komasa, W. Cencek, J. Rychlewski, Computational Methods in Science and Technology 2, 87 (1996), Scientific Publishers OWN, Poznań 1996
[11] F. Rolim, J. R. Mohallem, Acta Phys. Pol. 107, 661 (20050 and references therein
[12] H. Stachowiak, E. Boroński, Phys. Rev. B, 71, 245107 (2005)

E. A. G. Armour, J. Franz and J. Tennyson (eds.)
Explicitly Correlated Wavefunctions

Positron and positronium chemical physics by quantum Monte Carlo. Toward the exact calculation of scattering and annihilation properties

Massimo Mella

School of Chemistry, Cardiff University, Main Building, Park Place, Cardiff CF10 3AT (UK),
mellam@cf.ac.uk

I. INTRODUCTION

The accurate description of positron containing systems (e.g. molecules with a positron attached) represents a major methodological challenge for both Quantum Chemistry and Computational Physics due to the attractive Coulomb interaction between positrons (e^+) and electrons (e^-). This attraction induces a "pile-up" of the electron density at the coalescence point with the positron, an effect not present in a system containing only electrons. In the latter case, the probability of finding two electrons at coalescence is reduced by their mutual repulsion and by the Pauli principle, so that a single-particle description of the electronic structure (e.g. the Hartree-Fock wave function) is a suitable starting point for any improvement. Thus, it has been possible to devise effective methods correcting for the lack of a correlated description of the electronic motion, allowing the routine calculation of energy differences with an accuracy better than 1 kcal/mol (chemical accuracy) for systems containing 10-30 electrons.

Unfortunately, the different behaviour of the wave function for a positronic system close to the e^+-e^- coalescence point is such that the description provided by a single-particle model wave function gives a fairly inaccurate representation of the relative e^+-e^- distribution. As a consequence, this forces to use explicitly interparticle distances in the model wave function, with the net effect of making the evaluation of expectation values more difficult. In other words, one has to rely on complicate and expensive treatments even for small systems, a reason perhaps explaining the limited number of *ab initio* calculations published on positronic species so far. Among these, noteworthy are the work carried out by the Darwin group [1] on e^+A and PsA species (where A is an atom with either one or two valence electrons) and by Strasburger on small molecules [2].

Recently, quantum Monte Carlo (QMC) methods have emerged as an alternative to more standard *ab initio* techniques for the treatment of positronic systems [3–5]. The advantage of QMC, in its two variants variational Monte Carlo (VMC) [6] and diffusion Monte Carlo (DMC) [7], is provided by the possibility of estimating expectation values of explicitly correlated wave functions without the analytical calculation of complicate integrals. Perhaps even more interesting, it is the ability of DMC of projecting out excited state components from any model function, therefore sampling the ground state of the Schrödinger equation. This has allowed, so far, the accurate study of positronic systems containing up to 12 electrons, a quite demanding task for standard *ab initio* methods. With the computer resources available nowadays, one would also expect to be able to confront species containing up to 50 electrons with an accuracy better than the chemical one.

Bearing in mind the latter comments, this contribution focuses only on the most recent advancements of QMC tailored to describe positronic systems. In particular, we wish to describe two QMC-based schemes tackling the calculation of annihilation rates and the description of collisional events between a molecule (albeit small) and e^+ or a positronium atom (Ps).

II. ELASTIC SCATTERING FROM QMC SIMULATIONS

Obtaining accurate information on e^+ and Ps scattering off atoms and molecules requires an accurate description of the attractive interaction between the impinging projectile and the electrons of the target molecule, especially when the target has a large polarizability. The natural ability of QMC to recover a large fraction (≥90%) of the electron-electron correlation energy and completely the positron-electron one makes it a suitable candidate for this task. However, QMC is intrinsically a ground state method, a peculiarity that, at least in principle, limits its use in this field. Luckily, both e^+ and Ps seem to form complexes presenting a single bound state at most, therefore facilitating the task of maintaining scattering states orthogonal to bound ones. So, following suggestions made by Alhassid and Koonin [9] and by Carlson *et al.* [10], we implemented a QMC-based scattering method which closely mirrors the R-matrix approach [11] used in electron scattering [12] and recently extended to deal with projectile having an internal structure [13].

Briefly, the methodology requires to simulate the composite target-projectile system enclosed in a spherical box with rigid walls (thus preventing the projectile from leaving the interaction region) and it is well suited to deal with low energy scattering. Assuming the box radius to be sufficiently large, the total wave function near the boundary of the box can be written as product of three wave functions, two describing the internal structure of the fragments and one describing their relative motion. This allows us to compute the momentum k of the impinging projectile using the energy difference between the composite system, the target and the projectile constrained in the box. In turn, knowing k allows one to compute the phase shift for the partial wave under study.

As first application, the methodology was used to study the scattering of positronium (Ps) off H and He, the latter being a demanding case due to its internal structure. In both cases, several excited states for the composite systems were computed during a single simulation using the correlation function DMC approach [14]. This allows one to increase the magnitude of the relative momentum accessible by the simulations without violating the assumption of weak interaction near the boundary of the box. Figure 1 presents the results obtained for He using QMC and compares them with other recent theoretical results in which a rigid description of the He target was implemented.

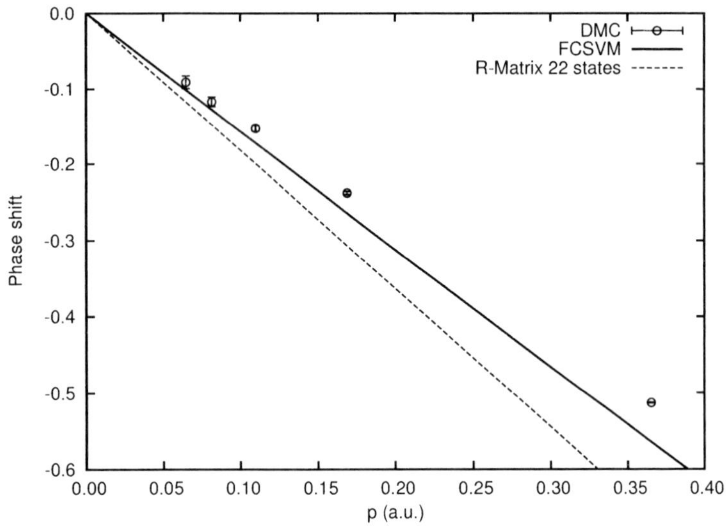

Figure 1: Phase shift as a function of the momentum of the impinging projectile for the scattering of Ps off He. The solid line (FCSVM) represents the results obtained using a stabilization method in conjunction with stochastic variational minimization (SVM), explicitly correlated Gaussians and a frozen (FC) description of the He target [15]. The dashed line indicates the results of a R-matrix calculation including 22 pseudo-states in the description of Ps and a frozen He target [13].

The results in Figure 1 seem to suggest that the "flexible" description provided by QMC for both fragments "softens" somewhat the target, substantially reducing the collisional cross section. In fact, the latter decreases from 13.2 a.u. and 10.56 a.u. for the R-matrix and the FCSVM results, respectively, to 7.89 a.u. in the QMC case.

III. ANNIHILATION RATES FROM QMC SIMULATIONS

The calculation of positron annihilation rates and Z_{eff} requires the estimate of the expectation value for the Dirac delta operator $\delta(r_{+-})$ (r_{+-} is the e^+-e^- distance). In principle, this expectation value may be obtained counting the number of QMC samples for which $r_{+-} \leq \epsilon$, where ϵ is small (e.g. 0.3 bohr) and *ad hoc*. More precisely, a pre-limit form of the Dirac delta is assumed, and the expectation value of this "discretized" operator is computed during the simulation. Unfortunately, this simple approach suffers from two important drawbacks, namely the divergence of its variance for $\epsilon \to 0$ and the small number of samples falling within the range $0 < r_{+-} \leq \epsilon$ (see Figure 2 for an example).

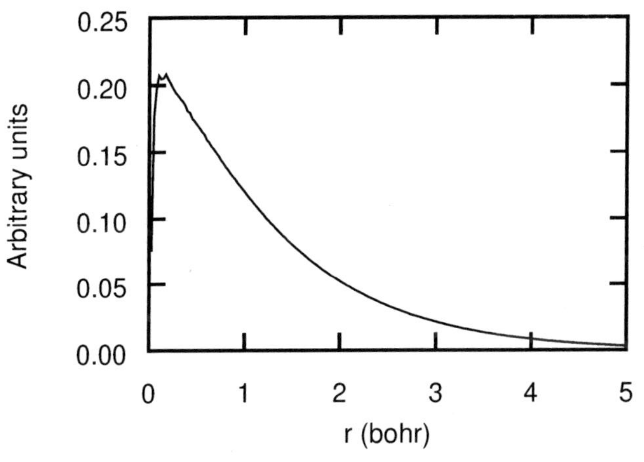

Figure 2: Electron-positron intracule distribution for a model system as computed using QMC simulations. Clearly visible in the picture, it is the lack of sampling in the region around the coalescence point.

A simple but effective way of circumventing these difficulties is provided by the analytical continuation of the e^+-e^- intracule distribution in the range of distance r_{+-} where the QMC sampling is scarce [8]. To do so, one may fit the intracule density in a region close to the e^+-e^- coalescence (say on the range 0.5-1.0 bohr in Figure 1) with an analytical function, which is then used to extrapolate to $r_{+-} \to 0$. The integral required to normalize the intracule density could then be split in two part, one computed using the fitted function and the other directly using the sampled distribution. We found convenient to carry out the extrapolation to $r_{+-} \to 0$ using a Jastrow factor of the form $J(r) = Ae^{-\frac{\alpha r+\beta r^2}{1+\gamma r}}$ with the exact cusp condition enforced by choosing the correct value for α. This introduce the correct limiting behaviour for the extrapolating function. Although approximate, this approach was found useful to study the behaviour of $\langle \delta(r_{+-}) \rangle$ as a function of the nuclear distance in a model e$^+$LiH system [8]. Besides, its extreme simplicity makes it easy to extend to larger systems.

In spite of the reasonable success of the method described above, it would nevertheless be advantageous to derive a QMC technique that is formally exact, at least in the limit of statistical nature of the QMC simulations. Such task would require to make no approximation on the relative position between an electron-positron pair, and hence to compute the exact value of the intracule distribution at coalescence. In this respect, it was recently noticed that this goal can be easily reached within the framework of VMC simulations [16] by simply rewriting the expectation value over the Dirac delta as

$$\langle \delta(r_{+-}) \rangle_T = \frac{\int \delta(r_{+-})\Psi_T^2(\mathbf{R})d\mathbf{R}}{\int \Psi_T^2(\mathbf{R})d\mathbf{R}} = \frac{\int \Phi_T^2(\mathbf{X})\chi(\mathbf{r}_{+-})d\mathbf{R}}{\int \Psi_T^2(\mathbf{R})d\mathbf{R}} \tag{1}$$

where $\mathbf{R} = (\mathbf{r_1}, \mathbf{r_2}, .., \mathbf{r_N}, \mathbf{r_+})$ is the position of the N system electrons and the positron, $\mathbf{X} = (\mathbf{r_1}, \mathbf{r_2}, .., \mathbf{r_N})$ is the position of only the N electrons, $\Psi_T(\mathbf{R})$ is an approximate wave function for the system, $\chi(r_{+-})$ is a 3D normalized function of r_{+-} and $\Phi_T(\mathbf{X}) = \int \delta(\mathbf{r}_{+-})\mathbf{\Psi_T}(\mathbf{R})d\mathbf{r}_+$. Apart from providing a practical and efficient way to estimate $\langle \delta(r_{+-}) \rangle_T$ during a VMC simulation, this idea also paved the way for the exact calculation of this expectation value. Indicating with $\Psi_0(\mathbf{R})$ the exact wave function for the system and defining $\Phi_0(\mathbf{X}) = \int \delta(\mathbf{r}_{+-})\mathbf{\Psi_0}(\mathbf{R})d\mathbf{r}_+$, $\langle \delta(r_{+-}) \rangle_0$ can be rewritten as

$$\begin{aligned}\langle \delta(r_{+-}) \rangle_0 &= \frac{\int \Phi_0^2(\mathbf{X})d\mathbf{X}}{\int \Psi_0^2(\mathbf{R})d\mathbf{R}} \\ &= \frac{\int \Phi_0^2(\mathbf{X})d\mathbf{X}}{\int \Phi_T^2(\mathbf{X})d\mathbf{X}}\frac{\int \Phi_T^2(\mathbf{X})d\mathbf{X}}{\int \Psi_T^2(\mathbf{R})d\mathbf{R}}\frac{\int \Psi_T^2(\mathbf{R})d\mathbf{R}}{\int \Psi_0^2(\mathbf{R})d\mathbf{R}} \\ &= \mathcal{U}\frac{\int \oplus_T^{\in}(\mathbf{X})d\mathbf{X}}{\int \ominus_T^{\in}(\mathbf{R})d\mathbf{R}}\mathcal{V}^{-\infty}\end{aligned} \tag{2}$$

were the central ratio of integrals is given by Equation 1. The remaining two ratios can be estimated using the descendant weighting approach [17], making sure that the simulations required to estimate $\mathcal{U}$ initially sample $\Phi_T^2(\mathbf{X})$.

This approach was found useful to estimate accurately $\langle\delta(r_{+-})\rangle$ for the bound state of Ps^- and PsH. Moreover, we found that it could be easily extended to estimate Z_{eff} during the elastic scattering of e^+ off small atomic targets, as shown in Figure 2 for H.

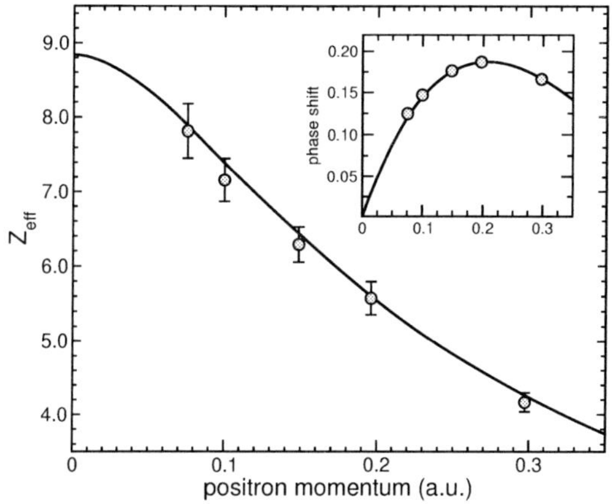

Figure 3: Z_{eff} as a function of the impinging positron momentum for the s-wave elastic scattering off H. The inset shows the phases shift values in the same range of momentum. The Z_{eff} estimate given by the solid line is taken from Ref. [18].

CONCLUSIONS AND OUTLOOK

Some recent advancements of QMC methodologies have been briefly presented. These advancements are tailored to tackle positronic systems, the latter not being easily amenable to more standard quantum chemistry treatments. It is the hope of the author that the applications succinctly presented in the previous paragraphs could spark more interest in this methodology and perhaps motivate further research in this area. Of course, the ideal goal would be to bring QMC to a stage in which it could provide some of the theoretical support needed in this field. Certainly, new phenomena just await to be unveiled by the synergic interplay between experiments and theory.

Acknowledgments

The author would like to thank Dario Bressanini, Simone Chiesa and Gabriele Morosi for the effort put into the development of a positronic version of QMC, and an EPSRC Advanced Research Fellowship (GR/R77803/01).

---

[1] J. Mitroy, M. W. J. Bromley and G. G. Ryzhikh, J. Phys. B: At. Mol. Opt. Phys. **35**, R81 (2002).
[2] K. Strasburger, J. Chem. Phys. **114** 615 (2001).
[3] D. M. Schrader, T. Ioshida and K. Iguchi, Phys. Rev. Lett. **68**, 3281 (1992); J. Chem. Phys. **98**, 7185 (1992).
[4] D. Bressanini, M. Mella and G. Morosi, J. Chem. Phys **108**, 4756 (1998); J. Chem. Phys **109**, 1716 (1998); J. Chem. Phys **109**, 5931 (1998).
[5] M. Mella, G. Morosi, D. Bressanini, J. Chem. Phys. **111**, 108 (1999); J. Chem. Phys. **112**, 1063 (2000); J. Chem. Phys. **114**, 10579 (2001).
[6] A. A. Frost, J. Chem. Phys. **41**, 478 (1964).
[7] J. B. Anderson, J. Chem. Phys. **60**, 1499 (1975).
[8] M. Mella, S. Chiesa, and G. Morosi, J. Chem. Phys. **116**, 2852 (2002).
[9] Y. Alhassid and S. E. Koonin, Ann. Phys. (N.Y.) **155**, 108 (1984).
[10] J. Carlson, V. R. Pandharipande and R. B. Wiringa, Nucl. Phys. A **424**, 47 (1984).
[11] E. P. Wigner and L. Eisenbud, Phys. Rev. **72**, 29 (1947).
[12] P. G. Burke, J. Tennyson, Mol. Phys. **103**, 2537 (2005).
[13] J. E. Blackwood, C. P. Campbell, M. T. McAlinden and H. R. J. Walters, Phys. Rev. A **60**, 4454 (1999).
[14] D. M. Ceperley and B. Bernu, J. Chem. Phys. **89**, 6316 (1988).
[15] J. Mitroy and I. A. Ivanov, Phys. Rev. A **65**, 012509 (2002).
[16] S. Chiesa, M. Mella and G. Morosi, Phys. Rev. A. **69**, 022701 (2004).
[17] M. Caffarel and P. Claverie, J. CHem. Phys. **88**, 1088 (1988).
[18] J. Mitroy and I. A. Ivanov, Phys. Rev. A **65**,042705 (2002).

E. A. G. Armour, J. Franz and J. Tennyson (eds.)
Explicitly Correlated Wavefunctions

Spin-unrestricted second-order Møller–Plesset correlation energies of the atoms He through Kr

S Höfener and W Klopper*

Lehrstuhl für Theoretische Chemie, Institut für Physikalische Chemie, Universität Karlsruhe (TH), D-76128 Karlsruhe, Germany

The second-order Møller–Plesset (MP2) correlation energies of the atoms He through Kr have been computed using the explicitly-correlated MP2-R12 approach, large basis sets of primitive Gaussians, and a spin-unrestricted Hartree–Fock (UHF) reference state. Inner-core electrons were not correlated. Accurate data exist in the literature for the alkaline earth metals, the noble gases, and zinc. Our results for the valence-shell second-order correlation energies of these elements agree with the literature data to within 0.5%, and we expect that the results for the other atoms are similarly accurate.

I. INTRODUCTION

With the aim of providing reference data for the construction and optimization of basis sets for correlated electronic-structure calculations, we have recently computed the spin-unrestricted valence-shell second-order Møller–Plesset (MP2) correlation energies of the atoms He through Kr using the explicitly-correlated MP2-R12 method [1,2]. Work on the fifth and higher groups of the periodic table of the elements is in progress.

In the following sections, we shall briefly present the MP2-R12 method for a spin-unrestricted Hartree–Fock (UHF) reference state (Section II), the basis sets used in the present work (Section III), and the results obtained for the atoms He–Kr (Section IV). An outlook on future work concludes the article.

---

*To whom correspondence should be addressed (klopper@chem-bio.uni-karlsruhe.de)

II. THE MP2-R12 METHOD

In the spin-unrestricted MP2-R12 method, we minimize the Hylleraas functional

$$H[u_{ij}] = \langle u_{ij}|\hat{f}_1 + \hat{f}_2 - \varepsilon_i - \varepsilon_j|u_{ij}\rangle + 2\langle u_{ij}|r_{12}^{-1}|ij\rangle, \tag{1}$$

where $|u_{ij}\rangle$ is the pair function that is variationally optimized and $|ij\rangle$ the two-electron determinant containing the two occupied canonical spin-orbitals φ_i and φ_j,

$$|ij\rangle = \tfrac{1}{\sqrt{2}}\{\varphi_i(\mathbf{x}_1)\varphi_j(\mathbf{x}_2) - \varphi_j(\mathbf{x}_1)\varphi_i(\mathbf{x}_2)\}. \tag{2}$$

All quantities are assumed to be real, $\hat{f}_1$ and $\hat{f}_2$ are the Fock operators for electrons 1 and 2, respectively, and ε_i and ε_j are the orbital energies of the canonical Hartree–Fock spin-orbitals φ_i and φ_j, respectively. The second-order correlation energy is obtained as a sum of pair energies,

$$E^{(2)} = \sum_{i<j} \varepsilon_{ij} = \sum_{i<j} H[u_{ij}]_{\mathrm{min}}, \tag{3}$$

where the sum runs over all (valence-shell) occupied orbitals that are included in the correlation treatment.

The MP2-R12 method used in the present work is based on the orbital-invariant formulation of MP2-R12 theory [2] and was applied as implemented in the DIRCCR12-OS program [3]. In this method, the pair function is written as $|u_{ij}\rangle = |v_{ij}\rangle + |w_{ij}\rangle$, and the Hylleraas functional (1) consists of a sum of two terms that are individually minimized,

$$H[u_{ij}] = H^{\mathrm{MP2}}[w_{ij}] + \langle v_{ij}|\hat{f}_1 + \hat{f}_2 - \varepsilon_i - \varepsilon_j|v_{ij}\rangle + 2\langle v_{ij}|r_{12}^{-1}|ij\rangle, \tag{4}$$

with

$$|v_{ij}\rangle = \sum_{k<l} c_{ij}^{kl}\hat{Q}_{12}r_{12}|kl\rangle = \sum_{k<l} c_{ij}^{kl}(1-\hat{P}_1)(1-\hat{P}_2)\, r_{12}|kl\rangle \tag{5}$$

and

$$|w_{ij}\rangle = \sum_{a<b} t_{ij}^{ab}|ab\rangle. \tag{6}$$

The sum over k, l runs over the occupied Hartree–Fock orbitals, the sum over a, b over the virtual orbitals, $\hat{P}_1$ and $\hat{P}_2$ are the projection operators for electrons 1 and 2, respectively, onto the orbital space spanned by the one-electron basis set, and c_{ij}^{kl} and t_{ij}^{ab} are the variational parameters (amplitudes) to be determined. If only the valence-shell orbitals i, j are correlated, then the sum over k, l is restricted to the same set of valence-shell occupied orbitals.

The minimum of $H^{\mathrm{MP2}}[w_{ij}]$ is the conventional MP2 energy, and the R12 correction to this energy is calculated by minimizing the R12 Hylleraas functional

$$\begin{aligned} H^{\mathrm{R12}}[v_{ij}] &= \langle v_{ij}|\hat{f}_1 + \hat{f}_2 - \varepsilon_i - \varepsilon_j|v_{ij}\rangle + 2\langle v_{ij}|r_{12}^{-1}|ij\rangle \\ &= \sum_{k<l}\sum_{m<n} c_{ij}^{kl} B_{kl,mn}^{ij} c_{ij}^{mn} + 2\sum_{k<l} c_{ij}^{kl} V_{kl}^{ij}, \end{aligned} \tag{7}$$

with

$$B_{kl,mn}^{ij} = \langle kl|r_{12}\hat{Q}_{12}(\hat{f}_1 + \hat{f}_2 - \varepsilon_i - \varepsilon_j)\hat{Q}_{12}r_{12}|mn\rangle \tag{8}$$

and

$$V_{kl}^{ij} = \langle kl|r_{12}\hat{Q}_{12}r_{12}^{-1}|ij\rangle. \tag{9}$$

The evaluation of the many-electron integrals occuring in Eqs. (8) and (6) has been described in detail elsewhere [4].

In the present work, we are concerned with MP2-R12 energies of open-shell atoms, using a spin-unrestricted Hartree–Fock reference state with α and β spin orbitals. Hence, we decompose the MP2-R12 energy as follows:

$$E^{(2)} = \sum_{i_\alpha<j_\alpha} \varepsilon_{i_\alpha,j_\alpha} + \sum_{i_\beta<j_\beta} \varepsilon_{i_\beta,j_\beta} + \sum_{i_\alpha,j_\beta} \varepsilon_{i_\alpha,j_\beta}, \tag{10}$$

where the sum over i_α, j_α runs over the (valence-shell) occupied spin-orbitals with α spin and the sum over i_β, j_β over those with β spin. The sums over k, l and m, n in Eq. (7) can be decomposed analogously, and thus, in the spin-unrestricted case, it remains to evaluate integrals of the types

$$V_{k_\alpha l_\alpha}^{i_\alpha j_\alpha} = \langle k_\alpha l_\alpha|r_{12}\hat{Q}_{12}r_{12}^{-1}|i_\alpha j_\alpha\rangle, \tag{11}$$

$$V_{k_\beta l_\beta}^{i_\beta j_\beta} = \langle k_\beta l_\beta|r_{12}\hat{Q}_{12}r_{12}^{-1}|i_\beta j_\beta\rangle, \tag{12}$$

$$V_{k_\alpha l_\beta}^{i_\alpha j_\beta} = \langle \varphi_{k_\alpha}\varphi_{l_\beta}|r_{12}\hat{Q}_{12}r_{12}^{-1}|\varphi_{i_\alpha}\varphi_{j_\beta}\rangle, \tag{13}$$

and

$$B_{k_\alpha l_\alpha,m_\alpha n_\alpha}^{i_\alpha j_\alpha} = \langle k_\alpha l_\alpha|r_{12}\hat{Q}_{12}(\hat{f}_1 + \hat{f}_2 - \varepsilon_{i_\alpha} - \varepsilon_{j_\alpha})\hat{Q}_{12}r_{12}|m_\alpha n_\alpha\rangle, \tag{14}$$

$$B_{k_\beta l_\beta,m_\beta n_\beta}^{i_\beta j_\beta} = \langle k_\beta l_\beta|r_{12}\hat{Q}_{12}(\hat{f}_1 + \hat{f}_2 - \varepsilon_{i_\beta} - \varepsilon_{j_\beta})\hat{Q}_{12}r_{12}|m_\beta n_\beta\rangle, \tag{15}$$

$$B_{k_\alpha l_\beta,m_\alpha n_\beta}^{i_\alpha j_\beta} = \langle \varphi_{k_\alpha}\varphi_{l_\beta}|r_{12}\hat{Q}_{12}(\hat{f}_1 + \hat{f}_2 - \varepsilon_{i_\alpha} - \varepsilon_{j_\beta})\hat{Q}_{12}r_{12}|\varphi_{m_\alpha}\varphi_{n_\beta}\rangle. \tag{16}$$

Note that simple orbital products such as $\varphi_{k_\alpha}\varphi_{l_\beta} \equiv \varphi_{k_\alpha}(\mathbf{x}_1)\varphi_{l_\beta}(\mathbf{x}_2)$ occur in Eqs. (13) and (16) for unlike spins, whereas two-electron determinants such as $|k_\alpha l_\alpha\rangle$ and $|k_\beta l_\beta\rangle$ occur in the like-spin $\alpha\alpha$ and $\beta\beta$ cases, respectively.

The DIRCCR12-OS program [3] was used to compute all of the matrices $\mathbf{B}^{ij}$ and vectors $\mathbf{V}^{ij}$ for the three spin cases $\alpha\alpha$, $\beta\beta$, and $\alpha\beta$ [5].

III. BASIS SETS

The basis sets used in the MP2-R12 calculations were derived mostly from the (augmented) correlation-consistent polarized-valence quintuple-zeta basis sets of Dunning and co-workers [6–10] and in a few rare cases from the def2-QZVPP basis of Weigend and Ahlrichs [11]. All of the basis sets used were fully uncontracted, and in general only a few tight polarization functions were added. The s-block elements and the atoms Ga–Kr required special attention, because we wanted to correlate more electrons than the original basis sets had been designed for. In the following, we briefly describe how we constructed our basis sets, which are available upon request.

He: Sets of tight functions (1d1f1g) with exponents $\zeta_{N+1} = \zeta_N^2/\zeta_{N-1}$ were added to the uncontracted aug-cc-pV5Z basis. Li, Be: Sets of diffuse functions (1s1p1d1f1g1h) as well as tight functions (1f1g1h) were added to the uncontracted cc-pV5Z basis. B–F: Sets of tight functions (2d1f1g1h) were added to the uncontracted aug-cc-pV5Z basis. Ne, Na: Basis sets of the types 15s9p10d7f6g3h and 21s12p9d9f6g5h, respectively, were constructed from the uncontracted def2-QZVPP basis [11]. Mg: Several functions were added to the uncontracted aug-cc-pV5Z basis to form a basis of the type 21s15p9d7f6g5h. Al–Ar: Sets of tight functions (1d1f1g1h) were added to the uncontracted aug-cc-pV(5+d)Z basis [8]. K, Ca: Basis sets of the types 30s21p12d9f8g7h and 29s21p14d9f8g7h, respectively, were constructed from the uncontracted def2-QZVPP basis [11]. Sc–Zn: Sets of tight functions (1f1g1h) were added to the uncontracted aug-cc-pV5Z basis of Balabanov and Peterson [10]. Ga–Kr: Basis sets of the type 29s20p14d9f8g7h were constructed from the uncontracted aug-cc-pV5Z basis of Wilson and co-workers [9].

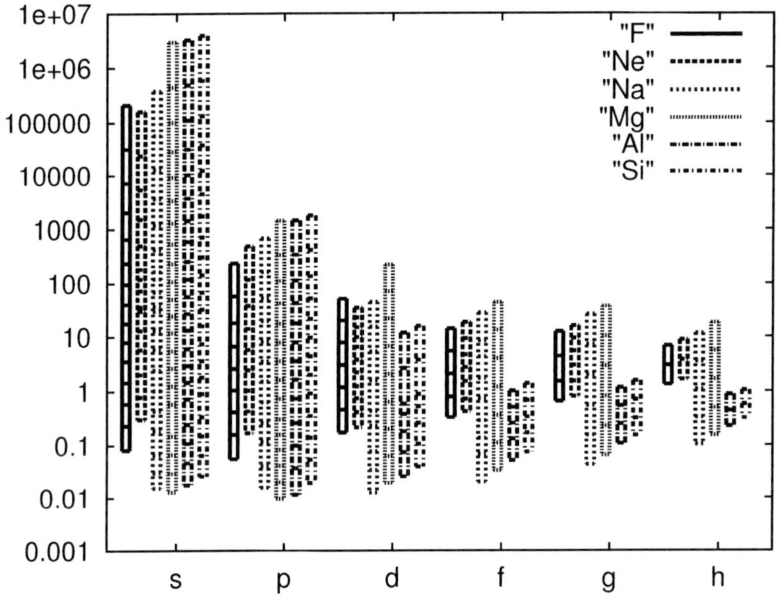

FIG. 1. Graphical representation of the exponents of the basis sets for the atoms F, Ne, Na, Mg, Al, and Si.

The basis sets for Ne and Na were derived from the def2-QZVPP basis, because this yielded a lower Hartree–Fock energy than the (aug-)cc-pV5Z basis. The def2-QZVPP basis was also chosen for K (for which no cc-pV5Z basis exists) and Ca. For the alkali and alkaline earth metals as well as the main group elements of the fourth period, we added extra functions in order to be able to correlate selected orbitals of the shell just below the outer-valence shell. For example, we correlated the 2s and 2p orbitals of Na and Mg and the 3d orbitals of Ga–Kr. In order to be able to do so, the exponents of the polarization functions had to cover a wide range of exponents, as depicted in Figure 1.

IV. RESULTS

The UHF and MP2-R12 energies of the atoms He–Kr are presented in Table 1, together with the term symbol of the atomic ground state (which does not necessarily represent the symmetry of the UHF solution) and the number of electrons that were correlated in the MP2-R12 approach. Data from the literature [12–16] for atoms with closed-shell $^1S_0$ ground states are given for comparison. The MP2-R12 energies consist of conventional MP2 energies and R12 corrections, and both contributions are given.

The R12 contributions amount to roughly 2–5% of the conventional MP2 energies. They are only larger than 5% for the s-block elements K (7%) and Ca (5.5%). For He and the elements of the second period, the R12 contributions are 2–3%, for the elements of the third period and the d-block of the fourth period, they amount to 2–4%, and for the main group elements Ga–Kr, they vary between 3.5 and 4.5%. These small percentages indicate that the MP2-R12 energies may be viewed as accurate estimates of the corresponding basis-set limit. For the closed-shell atoms, our MP2-R12 energies agree to within 0.5% with the literature data. Hence, as a rule of thumb, the large uncontracted basis sets used in the present work yield conventional MP2 energies of about 95% of the basis-set limit, whereas the MP2-R12 method yields about 99.5%. In other words, the R12 corrections reduce the basis-set error by one order of magnitude in these basis sets.

V. OUTLOOK

We plan to extend the present work to calculations of the MP2 energies of atoms from the fifth and higher groups of the periodic table of the elements. We feel that knowledge of the atomic basis-set limits might facilitate the optimization of basis sets for correlated calculations, particularly when developing balanced basis sets that recover roughly the same (prescribed) amount of correlation energy for all of the atoms in a molecule.

Furthermore, we plan to assess the performance of basis-set extrapolation techniques for spin-unrestricted MP2 energies. We shall not only investigate the two-point formula of Helgaker and co-workers [17], which exploits the X^{-3}-convergence of correlation energies with the cardinal number X, but also similar extrapolations using X^{-5} and X^{-3} formulas for like and unlike spins, respectively, as done in the complete-basis-set (CBS) approach of Petersson [18]. Such a scheme has already been explored for singlet and triplet pair energies of closed-shell systems in earlier work [19]. In this context, it also seems interesting to assess the performance of the spin-component-scaled MP2 approach (SCS-MP2) of Grimme [20], in which parallel- and antiparallel-spin pair-correlation energies are scaled separately (see also Ref. [21]).

Finally, the present results may serve as reference data for the construction and optimization of auxiliary basis sets for the RI-MP2-R12 method, in which 4-center-2-electron integrals are replaced by sums of products of 3-center-2-electron integrals by virtue of the resolution-of-the-identity (RI) approximation. This method was recently implemented [22] in the TURBOMOLE program, following previous work by Manby [23]. The results of the present work may help to investigate the requirements on the auxiliary basis set needed for the RI approximation in the RI-MP2-R12 method.

TABLE 1: Spin-unrestricted Hartree–Fock (UHF) and second-order Møller–Plesset correlation energies of the atoms He through Kr. All energies are given in mE_h.

Atom	Term	n_e^a	UHF	MP2	R12	MP2-R12	Lit.^{b}
He	$^1S_0$	2	−2 861.63	−36.71	−0.55	−37.26	−37.38^c
Li	$^2S_{1/2}$	3	−7 432.75	−40.57	−0.89	−41.46	
Be	$^1S_0$	4	−14 573.01	−74.51	−1.58	−76.09	−76.36^c
B	$^2P_{1/2}$	3	−24 533.13	−48.66	−0.97	−49.63	
C	$^3P_0$	4	−37 693.69	−74.94	−1.56	−76.50	
N	$^4S_{3/2}$	5	−54 404.47	−108.27	−2.20	−110.47	
O	$^3P_2$	6	−74 818.85	−166.02	−4.26	−170.28	
F	$^2P_{3/2}$	7	−99 416.09	−234.36	−6.42	−240.78	
Ne	$^1S_0$	8	−128 546.85	−312.22	−8.20	−320.42	−320.20^d
Na	$^2S_{1/2}$	9	−161 858.81	−309.58	−10.51	−320.09	
Mg	$^1S_0$	10	−199 614.61	−340.34	−13.17	−353.51	−354.58^e
Al	$^2P_{1/2}$	3	−241 880.77	−39.41	−0.98	−40.39	
Si	$^3P_0$	4	−288 858.85	−62.17	−1.61	−63.78	
P	$^4S_{3/2}$	5	−340 719.18	−92.60	−2.35	−94.95	
S	$^3P_2$	6	−397 513.26	−134.15	−4.54	−138.69	
Cl	$^2P_{3/2}$	7	−459 489.80	−185.49	−6.96	−192.45	
Ar	$^1S_0$	8	−526 817.35	−244.70	−9.58	−254.28	−254.57^d
K	$^2S_{1/2}$	9	−599 164.69	−259.34	−18.05	−277.39	
Ca	$^1S_0$	10	−676 758.00	−309.74	−16.91	−326.65	−326.93^f
Sc	$^2D_{3/2}$	3	−759 740.90	−30.74	−0.91	−31.65	
Ti	$^3F_2$	4	−848 414.70	−43.27	−1.16	−44.43	
V	$^4F_{3/2}$	5	−942 893.59	−65.06	−1.49	−66.55	
Cr	$^7S_3$	6	−1 043 356.78	−102.19	−1.16	−103.35	
Mn	$^6S_{5/2}$	7	−1 149 869.83	−133.07	−2.43	−135.50	
Fe	$^5D_4$	8	−1 262 455.53	−204.67	−6.30	−210.97	
Co	$^4F_{9/2}$	9	−1 381 427.44	−280.65	−10.31	−290.96	
Ni	$^3F_4$	10	−1 506 881.79	−369.66	−14.47	−384.13	
Cu	$^2S_{1/2}$	11	−1 638 964.23	−564.14	−20.61	−584.75	
Zn	$^1S_0$	12	−1 777 848.10	−577.22	−23.24	−600.46	−599.32^g
Ga	$^2P_{1/2}$	13	−1 923 264.49	−578.88	−20.84	−599.72	
Ge	$^3P_0$	14	−2 075 363.92	−594.83	−21.80	−616.63	
As	$^4S_{3/2}$	15	−2 234 239.81	−623.49	−21.94	−645.43	
Se	$^3P_2$	16	−2 399 875.46	−665.49	−26.35	−691.84	
Br	$^2P_{3/2}$	17	−2 572 448.48	−716.37	−30.00	−746.37	
Kr	$^1S_0$	18	−2 752 054.93	−773.67	−35.12	−808.79	−810.50^d

a Number of electrons included in the correlation treatment. b Reference values taken from the literature. c Ref. [12]. d Ref. [13]. e Ref. [14]. f Ref. [15]. g Ref. [16].

Acknowledgement

This research has been supported by the Deutsche Forschungsgemeinschaft through the Center for Functional Nanostructures (CFN, Project No. C3.3). It has been further supported by a grant from the Ministry of Science, Research and the Arts of Baden-Württemberg (Az: 7713.14-300). We thank David P. Tew for comments on the manuscript.

---

[1] Klopper W and Kutzelnigg W 1987 *Chem. Phys. Lett.* **134** 17
[2] Klopper W 1991 *Chem. Phys. Lett.* **186** 583
[3] Noga J, Klopper W, Helgaker T and Valiron P DIRCCR12-OS, an integral-direct explicitly-correlated coupled-cluster program, http://www-laog.obs.ujf-grenoble.fr/~valiron/ccr12
[4] Kutzelnigg W and Klopper W 1991 *J. Chem. Phys.* **94** 1985
[5] Noga J, Valiron P and Klopper W 2001 *J. Chem. Phys.* **115** 2022
[6] Dunning T H Jr 1989 *J. Chem. Phys.* **90** 1007
[7] Kendall R A, Dunning T H Jr and Harrison R J 1992 *J. Chem. Phys.* **96** 6796
[8] Woon D E and Dunning T H Jr 1993 *J. Chem. Phys.* **98** 1358
[9] Wilson A K, Woon D E, Peterson K A and Dunning T H Jr 1999 *J. Chem. Phys.* **110** 7667
[10] Balabanov N B and Peterson K A 2005 *J. Chem. Phys.* **123** 064107
[11] Weigend F and Ahlrichs R 2005 *Phys. Chem. Chem. Phys.* **7** 3297
[12] Bukowski R, Jeziorski B and Szalewicz K 1999 *J. Chem. Phys.* **110** 4165
[13] Flores J R 1993 *J. Chem. Phys.* **98** 5642
[14] Flores J R and Redondo P 1993 *Int. J. Quantum Chem.* **45** 563
[15] Flores J R 1994 *J. Comput. Chem.* **15** 782
[16] Flores J R and Redondo P 1993 *J. Phys. B: At. Mol. Opt. Phys.* **26** 2251
[17] Helgaker T, Klopper W, Koch H and Noga J 1997 *J. Chem. Phys.* **106** 9639
[18] Petersson G A, Yee A K and Bennett A 1985 *J. Chem. Phys.* **83** 5105
[19] Klopper W 2001 *Mol. Phys.* **99** 481
[20] Grimme S 2003 *J. Chem. Phys.* **118** 9095
[21] Jung Y S, Lochan R C, Dutoi A D and Head-Gordon M 2004 *J. Chem. Phys.* **121** 9793
[22] Villani C and Klopper W 2005 *J. Phys. B: At. Mol. Opt. Phys.* **38** 2555
[23] Manby F R 2003 *J. Chem. Phys.* **119** 4607

E. A. G. Armour, J. Franz and J. Tennyson (eds.)
Explicitly Correlated Wavefunctions

Approximations in Linear R12 Theories

Edward F. Valeev [1,2]

[1] *School of Chemistry and Biochemistry, Georgia Institute of Technology, Atlanta, GA 30332-0400 USA*

[2] *Computational Chemical Sciences Group, Oak Ridge National Laboratory, Oak Ridge, TN 37831-6367, USA*

I. APPROXIMATIONS IN LINEAR R12 THEORIES

Linear R12-type methods [1] have emerged as one of the most practical electronic structure approaches which includes correlation between electrons explicitly into account. The technical prowess of these methods is due to the ingenious use of the resolution of the identity (RI) to approximate three- and higher-electron integrals which appear in explicitly correlated methods. Since only two-electron integrals appear, linear R12 methods of the Møller-Plesset second-order method can be applied to the same molecular systems as the conventional methods. Further improvements in computational efficiency are possible due to improved RI techniques [2–4], density fitting [22,6], and local correlation [7,8]. Ultimately, the coupling of linear R12-type methodology with advanced many-electron correlation methods, such as multireference perturbation theory, configuration interaction [9], and coupled-cluster method [10], will enable robust first-principles computation of chemically-accurate reaction enthalpies and rates.

One of the main barriers to wider adoption of linear R12 type methods is the use of numerous approximations in addition to RI. Many combinations of the approximations have been used, which complicates comparison of published results and advancement of the methodology. Here I will review the "other" approximations of linear R12-type methods and compare their effect on the total energy.

The linear R12-type ansatz for the first-order pair function,

$$|\tilde{\psi}_{ij}^{(1)}\rangle = \hat{v}_{ij}|ij\rangle + \hat{w}_{ij}|ij\rangle \quad (1)$$

$$\hat{v}_{ij} = t_{ab}^{ij}|ab\rangle\langle ij| \quad (2)$$

$$\hat{w}_{ij} = \hat{Q}_{12}c_{kl}^{ij}|kl\rangle f_{12}\langle ij| \quad (3)$$

augments the conventional contribution from products of unoccupied Hartree-Fock (virtual) orbitals, Eq. (2), with terms linear in the correlation factor, $f_{12} \equiv f(r_{12})$. Projector $\hat{Q}_{12}$ ensures orthogonality of the explicitly correlated terms to all products of orbitals in the Hartree-Fock basis. The optimal choice of the projector is referred to as Ansatz **2** [2] and can be written in several equivalent forms, e.g.,

$$\hat{Q}_{12} = (1 - \hat{O}_1)(1 - \hat{O}_2) - \hat{V}_1 \hat{V}_2 \quad , \tag{4}$$

where $\hat{O}$ and $\hat{V}$ are the projectors on the occupied and virtual orbitals, respectively.

For simplicity, only evaluation of MP2-R12 energy will be considered here. Due to imposed strong orthogonality, Hylleraas functional for the second-order energy decomposes into sum of pair contributions

$$F^{(2)}(\tilde{\psi}_{ij}^{(1)}) = 2\,\mathrm{Re}\,\langle ij|r_{12}^{-1}|\tilde{\phi}_{ij}^{(1)}\rangle + \langle\tilde{\phi}_{ij}^{(1)}|\hat{F}_1 + \hat{F}_2 - \epsilon_i - \epsilon_j|\tilde{\phi}_{ij}^{(1)}\rangle \quad . \tag{5}$$

where $\hat{F}$ and ϵ are the Fock operator and orbital energy respectively. It is a standard practice to define the following "intermediate" quantities which appear upon substitution of Eq. (1) into the Hylleraas functional: first-order Hamiltonian,

$$V_{kl}^{ij} = \langle kl|r_{12}^{-1}\hat{Q}_{12} f_{12}|ij\rangle \quad , \tag{6}$$

overlap,

$$X_{kl}^{ij} = \langle kl|f_{12}\hat{Q}_{12} f_{12}|ij\rangle \quad , \tag{7}$$

and normal,

$$B_{kl}^{ij} = \langle kl|f_{12}\hat{Q}_{12}(\hat{F}_1 + \hat{F}_2)\hat{Q}_{12} f_{12}|ij\rangle \quad , \tag{8}$$

and coupling,

$$A_{ab}^{ij} = \langle ab|(\hat{F}_1 + \hat{F}_2)\hat{Q}_{12} f_{12}|ij\rangle \quad , \tag{9}$$

zeroth-order Hamiltonian. No assumptions, other than RI, are necessary to evaluate the first two intermediates. The latter two, however, are usually computed using many additional approximations.

The first approximation is to assume Extended Brillouin condition (EBC) [11],

$$(1 - \hat{P})\hat{F}\hat{V} \overset{\mathrm{EBC}}{\approx} 0 \quad , \tag{10}$$

where $\hat{P}$ is the projector on all Hartree-Fock orbitals. EBC holds exactly if the Hartree-Fock basis is complete. Because early linear R12 computations required a large Hartree-Fock basis set to use in the resolution of the identity, the use of EBC was perfectly logical. Technical advantages due to the assumption of EBC are significant: 1) the coupling matrix vanishes; 2) expressions for matrix element of B become simpler. Klopper and Samson were first to investigate MP2-R12 methods which did not assume EBC, however, the effect of EBC on the energy was not reported [2]. May, Valeev, Polly, and Manby concluded that the effect of EBC was much smaller than the remaining basis set incompleteness error in the MP2-R12 energy, but comparable to the error in the energies computed using a much more efficient Slater-type geminal correlation factor [12]. Their recommendation was therefore to avoid using EBC in linear R12-type calculations. A more rigorous recent study by Valeev [13] established that the effect of EBC varies between +0.5 and -2.0 %, depending on the correlation factor, when aug-cc-pVDZ Hartree-Fock basis is used. The EBC effect is significantly less than 0.5% when aug-cc-pVTZ basis is used.

Generalized Brillouin condition (GBC) [11],

$$(1-\hat{P})\hat{F}\hat{O} \overset{\text{GBC}}{\approx} 0 \quad , \tag{11}$$

also becomes exact when the Hartree-Fock basis is complete. Compared to EBC, GBC should be easier to satisfy in practice. In atomic computations, for example, GBC is exact if the Hartree-Fock basis is saturated up to the maximum angular momentum of the occupied orbitals. The use of GBC only affects the matrix elements of B, but simplifies them substantially. May *et al.* were first to examine how assumption of GBC affected MP2-R12 energies [12]. They concluded that it had negligible effect on the energies. A recent study by Valeev found that the effect of GBC is less than 0.1% and significantly smaller for bounded correlation factors [13].

To consider the last approximation, labeled as $[\hat{K}_1+\hat{K}_2, f_{12}] \approx 0$, let us examine the standard approach to computation of matrix B [11]:

$$\begin{aligned} B^{ij}_{kl} &= \langle kl|f_{12}\hat{Q}_{12}(\hat{F}_1+\hat{F}_2)\hat{Q}_{12}f_{12}|ij\rangle \\ &= \langle kl|f_{12}\hat{Q}_{12}[(\hat{F}_1+\hat{F}_2),\hat{Q}_{12}f_{12}]|ij\rangle + \langle kl|f_{12}\hat{Q}_{12}f_{12}(\hat{F}_1+\hat{F}_2)|ij\rangle \\ &= \langle kl|f_{12}\hat{Q}_{12}[(\hat{F}_1+\hat{F}_2),\hat{Q}_{12}]f_{12}|ij\rangle + \langle kl|f_{12}\hat{Q}_{12}[(\hat{F}_1+\hat{F}_2),f_{12}]|ij\rangle \\ &\quad + \langle kl|f_{12}\hat{Q}_{12}f_{12}(\hat{F}_1+\hat{F}_2)|ij\rangle \quad . \end{aligned} \tag{12}$$

The first and third terms can be evaluated explicitly via RI insertions but they simplify significantly from assumptions of EBC/GBC and GBC, respectively. The second term has the same form as matrix elements of V and X (Eqs. (6) and (7)), but the matrix elements of the $[\hat{F}_1+\hat{F}_2, f_{12}]$ operator cannot be evaluated in closed form. The solution is to separate the kinetic energy, $\hat{T}$, and exchange operator, $\hat{K}$:

$$[\hat{F}_1 + \hat{F}_2, f_{12}] = [\hat{T}_1 + \hat{T}_2, f_{12}] - [\hat{K}_1 + \hat{K}_2, f_{12}] \tag{13}$$

Matrix elements of the first term can be computed in closed form, whereas the second term can be evaluated via RI insertion (standard approximation B) or dropped altogether (standard approximation A). The omission of the exchange commutator therefore leads to significant technical simplification of expressions for matrix elements B, but at the cost of slower asymptotic convergence than the rigorous method [11,2]. The effect of the $[\hat{K}_1 + \hat{K}_2, f_{12}] \approx 0$ assumption is to make the MP2-R12 energy more negative. The extent of the energy lowering is highly basis-set and correlation-factor dependent. The use of partial wave basis sets always seems to cause the MP2-R12/A energy to overshoot the limiting value [2]. When a correlation-consistent Hartree-Fock basis set is used, MP2-R12/B energies are too high and MP2-R12/A energies are usually closer to the basis set limit due to the fortuitous cancellation of errors. When a more flexible correlation factor is used, the error of energies computed in standard approximation B is much smaller and it is possible for the MP2-R12/A energy to overshoot the limiting value, by as much as 5 % at the aug-cc-pVDZ level [13].

A promising approach to avoid the use of any approximations other than RI is the approach of Kedžuch *et al.*, which they dubbed standard approximation C [14]. The only difference between approximations B and C is the manner in which intermediate B is manipulated. Initial applications indicated that MP2-R12/C energies are always closer to the limit than their MP2-R12/B counterparts. This result, however, seems to be an artifact of using one basis set for the Hartree-Fock calculation and RI. Because technical expressions of standard approximation C are simpler than others, and $[\hat{T}_1 + \hat{T}_2, f_{12}]$ are no longer necessary, it is likely that approximation C will be the preferred method in the future.

The consensus at the time of writing is that the use of GBC is completely warranted, whereas the assumption of EBC and omission of the exchange commutator is possible when the Hartree-Fock basis is at least of a triple-zeta quality and not too high of accuracy is sought. In general, one should avoid using of the latter approximations.

Acknowledgments

I would like to thank Prof. Edward A. G. Armour, Prof. Jonathan Tennyson, and Dr. Jan Franz for organizing CCP2 Workshop on Explicitly Correlated Wavefunctions and inviting me to give a talk. My research is partially supported by the LDRD Program of Oak Ridge National Laboratory, managed by UT-Battelle, LLC, for the U.S. Department of Energy under Contract No. DE-AC05-00OR22725; and by the U.S. NSF grant CHE-0443564.

---

[1] W. Kutzelnigg, Theor. Chim. Acta **68**, 445 (1985).
[2] W. Klopper and C. C. M. Samson, J. Chem. Phys. **116**, 6397 (2002).
[3] E. F. Valeev, Chem. Phys. Lett. **395**, 190 (2004).
[4] S. Ten-no, J. Chem. Phys. **121**, 117 (2004).
[5] F. R. Manby, J. Chem. Phys. **119**, 4607 (2003).
[6] S. Ten-no and F. R. Manby, J. Chem. Phys. **119**, 5358 (2003).
[7] H.-J. Werner and F. R. Manby, J. Chem. Phys. **124**, 054114 (2006).
[8] F. R. Manby, H.-J. Werner, T. B. Adler, and A. J. May, J. Chem. Phys. **124**, 094103 (2006).
[9] R. J. Gdanitz, Chem. Phys. Lett. **210**, 253 (1993).
[10] J. Noga, W. Klopper, and W. Kutzelnigg, *Recent Advances in Coupled-Cluster Methods*, page 1 (World Scientific, Singapore, 1997).
[11] W. Kutzelnigg and W. Klopper, J. Chem. Phys. **94**, 1985 (1991).
[12] A. J. May, E. Valeev, R. Polly, and F. R. Manby, Phys. Chem. Chem. Phys. **7**, 2710 (2005).
[13] E. F. Valeev, Linear f12-type explicitly correlated theories with approximate hartree-fock orbitals, manuscript in preparation.
[14] S. Kedžuch, M. Milko, and J. Noga, Int. J. Quantum Chem. **105**, 929 (2005).

E. A. G. Armour, J. Franz and J. Tennyson (eds.)
Explicitly Correlated Wavefunctions

Local and density fitting approximations in MP2-F12 theory

Frederick R. Manby

Centre for Computational Chemistry, School of Chemistry
University of Bristol, Bristol BS8 1TS, UK

The past few years have seen numerous developments in the field of explicitly correlated electronic structure theory. These fall broadly into two categories, addressing issues of accuracy and of efficiency. On accuracy, two key conclusions emerge. Firstly, the original R12 framework, based on the generalized Brillouin condition and the resolution of the identity, remains perfectly sound; and in any case the former can be avoided and the latter improved. Secondly, the replacement of linear r_{12} with just about any short-ranged form f_{12} delivers great improvements in accuracy. In terms of efficiency, it has been shown that two well tested approaches — density fitting and local approximations — can be used successfully in MP2-F12 theory to reduce greatly the expense of moving to more accurate basis sets and to larger molecules.

I. INTRODUCTION

A. MP2-R12 wavefunctions

We consider the class of MP2 theories with first order pair functions built from the conventional virtual orbital products, augmented by a few explicitly correlated terms:

$$|u_{ij}>= T^{ij}_{ab}|ab> + t^{ij}_{kl}\hat{Q}_{12}f_{12}|kl> \,. \tag{1}$$

Summation over repeated dummy indices is assumed here and throughout, and the conventional choices are made for orbital indices: $ijkl$ for occupied; $abcd$ for virtual; and $pqrs$ for any. In the original work by Kutzelnigg and Klopper the explicit dependence on r_{12} was linear, ie $f_{12} = r_{12}$ [1–3]. The summation over kl in the second term was restricted to $kl = ij, ji$, but extending the summation over all occupied pairs is preferable as it leads to orbital invariance [4].

Second order correlation energies can then be obtained by minimizing the Hylleraas functional

$$< u_{ij}|\hat{f}_1 + \hat{f}_2 - \epsilon_i - \epsilon_j|u_{ij} > +2 < u_{ij}|r_{12}^{-1}|ij > \;\; \geq \; \epsilon_{ij}^{(2)} \tag{2}$$

with respect to the amplitudes T_{ab}^{ij} and t_{kl}^{ij}.

The decomposition of the MP2 energy into a sum over pair functions relies on the strong orthogonality of the first order pair functions to the occupied space [5]. Whereas in conventional MP2 this can be taken as read, in MP2-R12 theory the pair functions must be explicitly projected using a strong orthogonality operator $\hat{Q}_{12}$. This can take a variety of forms, but the most relevant to current thinking in the field is

$$\hat{Q}_{12} = (1 - \hat{O}_1)(1 - \hat{O}_2)(1 - \hat{V}_1\hat{V}_2) \tag{3}$$

where $\hat{O} = |i><i|$ and $\hat{V} = |a><a|$.

The well known trouble with theories based on explicitly correlated wavefunctions is the introduction of integrals over the coordinates of numerous particles: even at the level of MP2 theory, the ansatz of Eq. (1) introduces 3- and 4-electron integrals. These must at all costs be avoided if progress is to be made beyond systems of a few electrons.

B. Resolution of the identity

The key technology that is brought to bear in the R12 class of theories is the resolution of identity (RI), which, miraculously, breaks up many-electron integrals into sums of products of 2-electron integrals [3]:

$$\begin{aligned} < ijk|r_{12}r_{23}^{-1}|klm > &\approx < ijk|r_{12}\hat{P}_2'r_{23}^{-1}|klm > \\ &= < ijkp'|r_{12}r_{34}^{-1}|kp'ml > \\ &= < ij|r_{12}|kp' >< p'k|r_{12}^{-1}|lm > . \end{aligned} \tag{4}$$

Here the RI is performed in an auxiliary basis set [6], and, as indicated below, the errors associated with this approximation can be made small.

II. ACCURACY

A comprehensive review of the developments in terms of accuracy is completely unnecessary here, if only because they are well described by the other contributors to the present volume. However, some comments may be in order to highlight the key findings of the past few years. First the facts: MP2-R12 performs very well at recovering the last minute details of correlation energy in large basis sets. It does not perform well in basis sets that are currently deemed reasonable for the treatment of large molecules, ie basis sets of double-, triple- or maybe quadruple-zeta quality.

Two reasons could be identified for this shortcoming: either the larger basis sets reduce the impact of certain approximations in R12 theory (for example the generalized Brillouin condition (GBC), or RI in the AO basis); or perhaps in the larger basis sets the R12 correction is so minute that any sources of error have a correspondingly diminished impact. In any case there was clearly something of a mystery to unravel, and it is now satisfying that a clear picture has emerged as to role of all of the various approximations in the theory.

Three sources of error are important in MP2-R12 theory:

1. Errors in the many-electron integrals
2. Approximations to simplify the Hylleraas functional
3. The form of the correlation factor $f_{12} = r_{12}$

These will be treated (briefly) in turn.

A. Many-electron integrals

Before 2002 there were two basic strategies for evaluating the many-electron integrals in explicitly correlated electronic structure theory: you could evaluate them analytically (possible for atoms, or for molecules if Gaussians are used [7,8]); or one could use the RI approximation in the AO basis, as proposed by Kutzelnigg [1] and developed by Klopper and Kutzelnigg [3]. The first is hideously expensive, and the second has the disadvantage of requiring large AO basis sets. It also makes it hard to distinguish the RI error from other sources of error in the theory.

The development of the RI in an auxiliary basis set is distinguished as being the first, the most important and the most obvious extension of the method [6]. It allows the RI approximation to be converged to high accuracy with all other aspects of the theory fixed. Improvements that build on this key step have included combinations of RI and density fitting methods [9], numerical quadrature (ie RI in a basis of grid points) [10], and RIs in auxiliary basis sets augmented by the given AO basis [11]. In any case suffice it to say that (with the possible exception of calculations involving heavy elements) the RI can be converged within chemical accuracy routinely for all of the integrals required in R12 theories.

B. Other approximations

The approximations in question are the generalized Brillouin condition (GBC, the assumption that the occupied orbitals exactly satisfy the integro-differential Fock equations); the extended Brillouin condition (EBC, the assumption that this holds for the virtuals); and the neglect of commutators of the form $[\hat{K}_1, r_{12}]$. None of these approximations is necessary [12,13], and the EBC and the neglect of $[\hat{K}_1, r_{12}]$ cannot be justified. These approximations in modest basis sets lead to errors too large to allow convergence to chemical accuracy [14]. The impact of the GBC, however, is small [14], and although it can now be avoided [12,13], it is satisfying to note that the original framework of the GBC coupled with RI approximations holds up even when small AO basis sets are used.

C. Form of the correlation factor

The biggest source of error in MP2-R12 is the use of the linear correlation factor r_{12} [14]. Even a single Gaussian function $\exp(-\alpha r_{12}^2)$ performs better for a wide range of exponents α, which is intriguing because a Gaussian does not describe the cusp in the wavefunction at all — rather it appears that the important thing is to describe the correlation hole at intermediate distances. A single Slater function $\exp(-\alpha r_{12})$ seems to be outstanding [15,14,16]. This was explicitly demonstrated by Valeev who showed that the results obtained using a single Slater function for the correlation factor were very close to those obtained by fully optimizing the first order pair function with an expansion in Gaussian geminals [17].

The fact that the necessary integrals over $\exp(-\alpha r_{12})$ can be evaluated using standard recurrence relations [15] (as can integrals over any kernel for which there is a Gaussian transform [18]) means that there is no longer any reason for programs to use the linear form of R12. Methods that use an alternative correlation factor are referred to as MP2-F12 [18].

III. EFFICIENCY

Explicitly correlated methods share the poor scaling with respect to basis set size and molecular size that restrict conventional methods to molecules of perhaps 10–20 atoms. For larger systems and higher efficiency it has turned out to be possible to apply two successful approximations — density fitting and local approximations — to MP2-R12 and -F12 theory. The former addresses the cost as the basis size per atom increases, reducing the scaling from quartic to cubic, while also reducing prefactors dramatically. The latter addresses scaling with repect to system size, producing, in the limit of large systems, methods that have low (potentially linear) scaling with respect to system size.

A. Density fitting

Ab initio treatments of electron correlation require the evaluation of electron repulsion integrals (ERIs)

$$< pr|r_{12}^{-1}|qs >= (pq|rs) = \int d\vec{r}_1 \int d\vec{r}_2 \, \frac{\psi_p^*(1)\psi_q(1)\psi_r^*(2)\psi_s(2)}{r_{12}} \tag{5}$$

and the cost of computing them (especially of the transformation from atomic to molecular orbitals), can rapidly become prohibitive. Several techniques are available for reducing the cost but density fitting (DF) has perhaps proven the most popular owing to it conceptual simplicity and to the extremely small and regular errors that it introduces to energies and other properties.

The idea behind DF is simple: the objects $|pq)$ and $|rs)$ that appear in the ERI are orbital product densities, and these can be expanded (approximately) in a basis set:

$$|pq) \approx |\widetilde{pq}) = D_A^{pq}|A). \tag{6}$$

The coefficients D_A^{pq} should be optimized by making the error $|pq) - |\widetilde{pq})$ small. One can then approximate the ERI by

$$(pq|rs) \approx (\widetilde{pq}|\widetilde{rs}) = D_A^{pq} J_{AB} D_B^{rs}, \tag{7}$$

which only involves 2- and 3-index objects on the right hand side, and in which

$$J_{AB} = \int d\vec{r}_1 \int d\vec{r}_2 \, \frac{A(1)B(2)}{r_{12}}. \tag{8}$$

This basic set-up is old: in 1959 Boys and Shavitt used it to avoid the tricky 3-centre Slater integrals in calculations on H_3 [19]. It saw considerable early application in DFT [20] in which context the whole density ρ is expanded in a basis set, to facilitate the evaluation of the Coulomb contribution to the Fock matrix. The key to the modern success of DF hangs on the choice of the criterion (or metric) by which the fitting residual $|pq) - |\widetilde{pq})$ is measured. In the earlier work this was simply taken as the self-overlap of the fitting residual, but in 1973 Whitten introduced the idea of using the Coulomb self-energy of the residual $(pq - \widetilde{pq}|pq - \widetilde{pq})$ as the fitting criterion. This has a dramatic effect on the accuracy of the method: using the simple overlap criterion, errors linear in the fitting residual enter into the expressions for the fitted integrals. Using the Coulomb criterion, however, the errors in the integrals are quadratic in the fitting residual, and the method can deliver very high accuracy using fitting basis sets only a few times larger than the AO basis. This was later elucidated by Dunlap, who realized that minimizing the Coulomb norm is equivalent to least squares fitting the electric fields [21].

When the Coulomb criterion is used, one minimizes

$$\Delta_{pq} = (pq - \widetilde{pq}|pq - \widetilde{pq}) \tag{9}$$

to find

$$D_A^{pq} = [\mathbf{J}^{-1}]_{\mathbf{AB}}\mathbf{J}_{\mathbf{B}}^{\mathbf{pq}} = \bar{\mathbf{J}}_{\mathbf{A}}^{\mathbf{pq}} \tag{10}$$

where

$$J_A^{pq} = \int d\vec{r}_1 \int d\vec{r}_2 \, \frac{A(1)\psi_p^*(2)\psi_q(2)}{r_{12}}. \tag{11}$$

The error in the fitted integral can be written

$$(pq|rs) - (\widetilde{pq}|\widetilde{rs}) = (pq - \widetilde{pq}|rs - \widetilde{rs}) + (\widetilde{pq}|rs - \widetilde{rs}) + (pq - \widetilde{pq}|\widetilde{rs}) \tag{12}$$

and the last two terms vanish (they are simply the residuals of the equations that were solved to obtain the coefficients), leaving

$$(pq|rs) - (\widetilde{pq}|\widetilde{rs}) = (pq - \widetilde{pq}|rs - \widetilde{rs}), \tag{13}$$

which is quadratic in the fitting error. Such fitting expression are called *robust* by Dunlap.

The cancellation of the linear terms happens only when the fitting criterion and the target integral type coincide. If density fitting is to be used for other types of integrals — such as those that appear in MP2-R12 or -F12 theory, the linear terms have to be subtracted explicitly. For integrals over some arbitrary kernel f_{12} one uses [22,18]

$$(ip|f_{12}|jq) = \bar{J}_{ip}^A F_{jq}^A + F_{ip}^A \bar{J}_{jq}^A - \bar{J}_{ip}^A F_{AB} \bar{J}_{jq}^B, \tag{14}$$

and for the kinetic energy commutator integrals the expression is only slightly more complicated:

$$(ip|[\hat{t}_1, f_{12}]|jq) = \bar{Y}_{ip}^A F_{jq}^A + X_{ip}^A \bar{J}_{jq}^A - \bar{Y}_{ip}^A F_{AB} \bar{J}_{jq}^B. \tag{15}$$

The various 2- and 3-index integrals required are given by

$$\begin{array}{llll} F_{AB} = (A|f_{12}|B) & F_{ip}^{A} = (A|f_{12}|ip) \\ X_{ip}^{A} = (A|[\hat{t}_1, f_{12}]|ip) & Y_{ip}^{A} = (A|[\hat{t}_1, r_{12}^{-1}]|ip) \end{array} \tag{16}$$

and their evaluation is discussed elsewhere for $f_{12} = r_{12}$ [22] and for an f_{12} expanded in Gaussians [18].

To summarize density fitting in MP2-R12 and MP2-F12 theory, we note that the very substantial gains in efficiency introduces practically no error [22]. Providing robust density fitting formulae are used for all target integral types, the errors introduced using conventional MP2 density fitting sets are completely negligible on the scale of chemical accuracy.

Having said that, a small error does creep into the approach when r_{12} is used in extended systems. This is merely a symptom of the poor numerics that arise when r_{12} becomes large, and disappears when short-range correlation factors are used. A complete explanation of this problem is given elsewhere [23].

B. Local approximations

The steep scaling of post-Hartree-Fock electron correlation methods with system size is now well known to be unphysical. In insulators the correlation phenomenon is dominated in the long range by dispersion, which decays as R^{-6} where R is the distance between two fragments of matter. Indeed, the high scaling of the correlation methods arises simply from the non-local nature of the canonical Hartree-Fock (HF) orbitals. Since the occupied orbitals can be rotated amongst themselves without changing the HF wavefunction, it is perfectly possible to choose a rotation that maximizes the locality of the orbitals. There are various criteria to define what is meant by 'locality' [24,25].

Once one has a basis of local orbitals (LMOs), it becomes possible to think in terms of the distance between orbital pairs, and thus to decide the level of theory at which to treat each pair based on their spatial separation. By introducing localized sets of virtual orbitals (typically projected atomic orbitals — PAOs), it becomes possible to restrict excitations from a given local orbital to virtual orbitals that are nearby. These ideas were developed by Pulay and Saebø [26–30] and realized as low-scaling implementations of MP2 and coupled cluster theory by Werner and Schütz [31–39].

In local MP2 (LMP2) theory the usual expansion of the first order wavefunction

$$|1> = T_{ab}^{ij}|\Psi_{ij}^{ab}> \tag{17}$$

is replaced by a similar expression with the summation ranges restricted

$$|1>=\sum_{i,j\in\mathcal{P}}\sum_{a,b\in[ij]}T^{ij}_{ab}|\Psi^{ab}_{ij}>. \tag{18}$$

Here $\mathcal{P}$ is the pair list, which contains those pairs of LMOs that lie close to each other (according to some criterion that can be varied). The orbital excitation domains $[ij]$ are sets of PAOs that lie sufficiently close to either of the LMOs i or j. The number of terms in the expansion in Eq. (17) grows like $\mathcal{O}(\mathcal{N}^{\triangle})$ as the system increases, whereas that in Eq. (18) grows linearly with system size.

In local MP2-F12 theory [23,40] the idea is to perform conventional MP2 for all (or most) pairs, and only to compute the F12 correction for strongly correlated pairs of electrons. This results in a method vastly more efficient than the conventional MP2-F12 theory.

In the work that has been done so far, the explicitly correlated ansatz has been restricted to MP2-F12/**2***A′. This assumes the EBC which allows the computational convenience of completely decoupling the F12 correction from the conventional MP2 part. As noted above, though, the EBC does introduce significant errors, and an important goal for future development in a local MP2-F12 theory with fewer approximations.

In conventional MP2-F12/**2***A′ the F12 part of the pair function has the form

$$|u_{ij}>=\hat{Q}_{12}f_{12}\sum_{kl}t^{ij}_{kl}|kl>. \tag{19}$$

In the local variant (snappily called MP2-F12/**2***A(loc)) the F12 part takes the form

$$|u_{ij}>=\hat{Q}_{12}f_{12}\sum_{kl\in\{ij,ji\}}t^{ij}_{kl}|kl>. \tag{20}$$

As noted previously, restricting the summation to $kl\in\{ij,ji\}$ breaks orbital invariance, but since local methods are inherently non-orbital-invariant this probably does not cause significant problems. Tests indicate that the errors introduced by this approximation are minor. In any case in the long term it would probably be better to increase the range of the summation over kl to include all pairs spatially close to ij.

Having made the ansatz above it becomes possible to localize other aspects of the theory. The RI can be restricted to include only functions in the spatially relevant region. In principle the projection operator itself can be localized. If density fitting is combined with local approximations (a good idea as this boosts efficiency enormously [23,40]) it is possible to localize the DF expansion as well.

The resulting method, now lumbered with the name DF-MP2-F12/**2***A(loc), is exciting. The accuracy in a given basis in the series aug-cc-pVnZ surpasses that of conventional MP2/aug-cc-pV(n + 2)Z, and the whole calculation (including HF) is cheaper than HF+MP2/aug-cc-pV(n + 1)Z. The efficiency is sufficient to make explicitly correlated calculations in aug-cc-pVTZ basis sets on enzyme active sites accessible. These are essentially at the basis set limit, and have been performed as part of a wider effort to perform chemically converged QM/MM calculations on enzymatic catalysis [41].

Further improvements in efficiency and accuracy are possible. Ours plans to achieve the former concentrate on improving the integrals technology to a level more comparable with state of the art Coulomb integral programs. For the latter we intend to implement the more accurate ansatz F12/**2**B; to include summations over locally determined domains of kl; and to implement the important developments in the RI expansion described by Valeev [11].

IV. CONCLUSIONS

This paper has described some recent developments in terms of accuracy and efficiency in explicitly correlated electronic structure theory. Regarding accuracy, the big surprise in the field has been that it is the choice of r_{12} in MP2-R12 theory, and not the use of the RI, that leads to the poor performance in smaller basis sets. Regarding efficiency, it has been possible to apply the highly successful combination of density fitting and local approximations with MP2-F12 theory. This leads to methods that can be used to obtain essentially converged MP2 correlation energies for large molecules (say 50 atoms) in a fairly routine manner.

Acknowledgements

The two people instrumental in the work I have presented are Andrew May (for the original work on MP2-F12 theory) and Hans-Joachim Werner (for all of the local work) and I am indebted to them. I am grateful to Edward Valeev, Wim Klopper and Seiichiro Ten-no for stimulating discussions. The research described has been funded by the EPSRC (GR/R93704/01) and the Royal Society.

[1] W. Kutzelnigg, Theor. Chim. Acta **68**, 445 (1985).

[2] W. Klopper and W. Kutzelnigg, Chem. Phys. Lett. **134**, 17 (1987).
[3] W. Kutzelnigg and W. Klopper, J. Chem. Phys. **94**, 1985 (1991).
[4] W. Klopper and J. Almlöf, J. Chem. Phys. **99**, 5167 (1993).
[5] O. Sinanoğlu, J. Chem. Phys. **36**, 706 (1962).
[6] W. Klopper and C. C. M. Samson, J. Chem. Phys. **116**, 6397 (2002).
[7] S. F. Boys, Proc. Roy. Soc. A **258**, 402 (1960).
[8] K. Singer, Proc. Roy. Soc. A **258**, 412 (1960).
[9] S. Ten-no and F. Manby, J. Chem. Phys. **119**, 5358 (2003).
[10] S. Ten-no, J. Chem. Phys. **121**, 117 (2004).
[11] E. F. Valeev, Chem. Phys. Lett. **395**, 190 (2004).
[12] E. F. Valeev, unpublished work (2006).
[13] S. Kedžuch, M. Milko, and J. Noga, Int. J. Quantum Chem. **105**, 929 (2005).
[14] A. J. May, E. Valeev, R. Polly, and F. R. Manby, Phys. Chem. Chem. Phys. **7**, 2710 (2005).
[15] S. Ten-no, Chem. Phys. Lett. **398**, 56 (2004).
[16] D. P. Tew and W. Klopper, J. Chem. Phys. **123**, 074101 (2005).
[17] E. F. Valeev, (2006), Presentation at CCP2 Workshop on Explicitly Correlated Wavefunctions, Nottingham.
[18] A. J. May and F. R. Manby, J. Chem. Phys. **121**, 4479 (2004).
[19] S. F. Boys and I. Shavitt, University of Wisconsin, Report WIS-AF-13 (1959).
[20] E. J. Baerends, R. H. Felton, and P. Roos, Chem. Phys. **2**, 41 (1973).
[21] B. I. Dunlap, J. W. D. Connoly, and J. R. Sabin, J. Chem. Phys. **71**, 3396 (1979).
[22] F. R. Manby, J. Chem. Phys. **119**, 4607 (2003).
[23] H.-J. Werner and F. R. Manby, J. Chem. Phys. **124**, 054114 (2006).
[24] S. F. Boys, in *Quantum Theory of Atoms, Molecules, and the Solid State*, edited by P. O. Löwdin, page 253 (Academic Press, New York, 1966).
[25] J. Pipek and P. G. Mezey, J. Chem. Phys. **90**, 4916 (1989).
[26] P. Pulay, Chem. Phys. Lett. **100**, 151 (1983).
[27] S. Saebø and P. Pulay, Chem. Phys. Lett. **113**, 13 (1985).
[28] P. Pulay and S. Saebø, Theor. Chim. Acta **69**, 357 (1986).
[29] S. Saebø and P. Pulay, J. Chem. Phys. **86**, 914 (1987).
[30] S. Saebø and P. Pulay, J. Chem. Phys. **88**, 1884 (1988).
[31] C. Hampel and H.-J. Werner, J. Chem. Phys. **104**, 6286 (1996).
[32] G. Hetzer, P. Pulay, and H.-J. Werner, Chem. Phys. Lett. **290**, 143 (1998).
[33] M. Schütz, G. Hetzer, and H.-J. Werner, J. Chem. Phys. **111**, 5691 (1999).
[34] G. Hetzer, M. Schütz, H. Stoll, and H.-J. Werner, J. Chem. Phys. **113**, 9443 (2000).
[35] M. Schütz and H.-J. Werner, Chem. Phys. Lett. **318**, 370 (2000).
[36] M. Schütz, J. Chem. Phys. **113**, 9986 (2000).
[37] M. Schütz and H.-J. Werner, J. Chem. Phys. **114**, 661 (2001).
[38] M. Schütz, J. Chem. Phys. **113**, 8772 (2002).
[39] M. Schütz, Phys. Chem. Chem. Phys. **4**, 3941 (2002).
[40] F. R. Manby, H.-J. Werner, T. B. Adler, and A. J. May, J. Chem. Phys. **124**, 094103 (2006).
[41] F. Claeyssens, J. N. Harvey, F. R. Manby, R. A. Mata, A. J. Mulholland, K. E.

Ranaghan, M. Schütz, S. Thiel, W. Thiel, and H.-J. Werner, Science (2006), submitted.

E. A. G. Armour, J. Franz and J. Tennyson (eds.)
Explicitly Correlated Wavefunctions

Beyond the Standard Approximation in Coupled Cluster R12 Theory

Jozef Noga[1,2] and Stanislav Kedžuch[2]

[1] *Department of Physical and Theoretical Chemistry, Faculty of Natural Sciences, Comenius University, SK-84215 Bratislava, Slovakia*
[2] *Institute of Inorganic Chemistry, Slovak Academy of Sciences, SK-84536 Bratislava, Slovakia*

I. INTRODUCTION

Explicitly correlated R12 coupled cluster theory (CC-R12) [1] has proven to be an efficient tool to achieve the basis set limit accuracy in calculations energies and properties of small molecules [2]. This theory resulted from a combination of the traditional coupled cluster ansatz by Čížek [4] and R12 ansatz by Kutzelnigg [5]. Current implementations are using the so called "standard approximation" (SA) to avoid computations of three and four-particle integrals [6]. This approximation assumes the finite spin orbital space being closed under the Fock operator, which means that not only the generalized Brillouin condition (GBC) but also the extended Brillouin condition (EBC) (vide infra) are valid. At the same time SA assumes that the resolution of identity can be well represented within the given computational basis set. The SA constraints imply that the basis has to be saturated at least at the level of $3L_{occ}$, where L_{occ} is the highest angular momentum involved in the occupied spin orbitals. Very strictly, even stronger condition may be required for CCSD-R12, ensuring that in addition the computational basis is saturated for each involved angular momentum [1].

Alternatively, one can use an auxiliary basis set for the resolution of identity with the aim to suppress the strong basis set requirement. This has been implemented recently within the MP2-R12 theory using diverse variants [7–9]. Nevertheless, the resulting formulas still contained integrals over the commutator of the kinetic operator with inter-electronic distance, whose computation represent the most demanding part of the whole MP2-R12 calculation. In our recent paper [10], we have suggested a formulation that is free from the pertinent integrals. At the same time we have reported results for a new variant (MP12-R12/C) that still used the standard approximation. Below, we shall briefly outline how the CC-R12 theory will change if the assumptions of the SA are abandoned. Preliminary results are reported for MP2-R12.

II. OUTLINE OF THE CC-R12 THEORY

In the following we will denote occupied orbitals as i, j, …, o; virtual spin orbitals within the computational basis as a, b, c; arbitrary spin orbitals within the computational basis as p, q, r, s. Greek letters κ, λ, μ, ν denote any spin orbitals within the complete basis, and α, β, γ, the spin orbitals from the complete basis that are outside the final computational one. The basis set is assumed to be orthonormal. We will use a tensor notation for the integrals over any n-body operator $\hat{o}_n$,

$$o^{\lambda_1 \ldots \lambda_n}_{\kappa_1 \ldots \kappa_n} = \langle \kappa_1 \ldots \kappa_n | \hat{o}_n | \lambda_1 \ldots \lambda_n \rangle, \tag{1}$$

as well as for integral products. Einstein summation convention is considered throughout, i. e. summations run over all indices that do not appear on both sides of the equations.

In general, the coupled cluster theory is based on the exponential ansatz for the wave function,

$$|\Psi\rangle = e^{\hat{S}} |\Phi\rangle, \tag{2}$$

where $|\Phi\rangle$ is a known reference function, if we restrict ourselves to a single-reference theory. In traditional theories, the operator $\hat{S}$ is identified by the electronic excitations from the occupied orbitals constituting the $|\Phi\rangle$, to certain (available) unoccupied orbitals, usually produced as a byproduct of the solution for $|\Phi\rangle$. Using the second quantization and normal ordering with respect to Φ the Hamiltonian is expressed as

$$\hat{H}_N = \hat{H} - \langle \Phi | \hat{H} | \Phi \rangle = \hat{F}_N + \hat{W}_N \,, \tag{3}$$

$$\hat{F}_N = f^{\lambda}_{\kappa} \tilde{a}^{\kappa}_{\lambda} = \left(h^{\lambda}_{\kappa} + \bar{g}^{\lambda i}_{\kappa i} \right) \tilde{a}^{\kappa}_{\lambda} \,, \tag{4}$$

$$\hat{W}_N = \tfrac{1}{4} \bar{g}^{\mu\nu}_{\kappa\lambda} \tilde{a}^{\kappa\lambda}_{\mu\nu} \,, \tag{5}$$

where $\hat{F}_N$ is the fock operator defined by the matrix elements f and replacement operators $\tilde{a}$. W_N similarly defines the two-electron part (g=r_{12}^{-1}). The conventional global excitation operator ($\hat{S} = \hat{T}$) is given as [4]:

$$\hat{T} = \sum_n \hat{T}_n \,; \qquad \hat{T}_n = \frac{1}{n!} \sum_{\substack{i_1,\ldots i_n \\ a_1 \ldots a_n}} t^{i_1,\ldots i_n}_{a_1 \ldots a_n} \tilde{a}^{a_1 \ldots a_n}_{i_1,\ldots i_n} \,, \tag{6}$$

where $\tilde{a}^{a_1 \ldots a_n}_{i_1,\ldots i_n}$ is a particular n-body excitation operator and t is the associated scalar amplitude.

In the explicitly correlated coupled cluster theory, the operator $\hat{S}$ constitutes from the conventional part $\hat{T}$ and a special operator $\hat{\mathcal{R}}$ that somehow introduces the inter-electronic coordinate (r_{12}) into the final wave-function. Having $\hat{S}{=}\hat{T}{+}\hat{\mathcal{R}}$, after multiplying the Schrödinger equation from the left by $\exp(-\hat{S})$ one arrives at:

$$e^{-\hat{T}}e^{-\hat{\mathcal{R}}}H_N e^{-\hat{\mathcal{R}}}e^{\hat{T}}|\Phi\rangle = e^{-\hat{T}}H_N^{\text{transcor}}e^{\hat{T}}|\Phi\rangle = E|\Phi\rangle. \tag{7}$$

In the last equation the similarity transformed Hamiltonian

$$H_N^{\text{transcor}} = e^{-\hat{\mathcal{R}}}H_N e^{-\hat{\mathcal{R}}} \tag{8}$$

resembles the idea of the transcorrelated Hamiltonian by Boys and Handy [3], though, they used a different operator as in the forthcoming.

As mentioned, the CC-R12 theory is related to the idea of Kutzelnigg [5] who showed that the electron-electron cusp is properly described if the wave function explicitly contains $r_{12}\Phi$. Having the global operator of inter-electronic distances $\hat{r}{=}\sum_{p>q} r_{pq}$ in normal ordered form

$$\hat{r}_N = \tfrac{1}{4}\bar{r}^{\kappa\lambda}_{\mu\nu}\tilde{a}^{\mu\nu}_{\kappa\lambda} + \bar{r}^{\kappa i}_{\mu i}\tilde{a}^{\mu}_{\kappa} \tag{9}$$

one gets

$$\hat{r}_N|\Phi\rangle = \left(\tfrac{1}{4}\bar{r}^{ij}_{ab}\tilde{a}^{ab}_{ij} + \tfrac{1}{4}\bar{r}^{ij}_{\alpha\beta}\tilde{a}^{\alpha\beta}_{ij} + \tfrac{1}{2}\bar{r}^{ij}_{a\beta}\tilde{a}^{a\beta}_{ij} + \bar{r}^{ij}_{aj}\tilde{a}^{a}_{i} + \bar{r}^{ij}_{\alpha j}\tilde{a}^{\alpha}_{i}\right)|\Phi\rangle\,. \tag{10}$$

A pseudo one-particle part is now clearly recognized. If parts that overlap with the traditional expansion of the wave function are outprojected one can introduce the operator $\mathcal{R}$ as:

$$\hat{\mathcal{R}} = \hat{\mathcal{R}}_1 + \hat{\mathcal{R}}_2 = c^i_k\tilde{\mathcal{R}}^k_i + \tfrac{1}{4}c^{ij}_{kl}\tilde{\mathcal{R}}^{kl}_{ij} \tag{11}$$

with

$$\tilde{\mathcal{R}}^k_i = \tilde{a}^k_i\bar{r}^{kj}_{\alpha j}\tilde{a}^{\alpha}_{k}; \qquad \tilde{\mathcal{R}}^{kl}_{ij} = \tilde{a}^{kl}_{ij}\left(\tfrac{1}{2}\bar{r}^{kl}_{\alpha\beta}\tilde{a}^{\alpha\beta}_{kl} + \bar{r}^{kl}_{a\beta}\tilde{a}^{a\beta}_{kl}\right). \tag{12}$$

A. The standard approximation

Within the standard approximation one assumes that the basis set is saturated such that matrix elements of $\hat{r}$ vanish as soon as just a single index belongs to the complementary basis (α). Equivalently, this means that the resolution of identity (RI) ($1 = |\kappa\rangle\langle\kappa|$) is describable in the given computational basis.

$$|\kappa\rangle\langle\kappa| \approx |p\rangle\langle p| \tag{13}$$

Consequently, $\hat{\mathcal{R}}$ is reduced to:

$$\hat{\mathcal{R}} \overset{\text{s.a.}}{=} \frac{1}{4} c_{kl}^{ij} \tilde{\mathcal{R}}_{ij}^{kl}; \qquad \tilde{\mathcal{R}}_{ij}^{kl} \overset{\text{s.a.}}{=} \frac{1}{2} \bar{r}_{\alpha\beta}^{kl} \tilde{a}_{kl}^{\alpha\beta} \tag{14}$$

Assuming the validity of GBC ($f_i^\alpha = 0$) and even EBC ($f_p^\alpha = 0$) is then natural and automatic. Finally, these approximations permit straightforward evaluation of matrix elements resulting in the CC-R12 theory [1]. As aforementioned, the price to pay is compulsory using of fairly extended basis sets.

B. Beyond the standard approximation

Let us exclude the exact treatment with three- and four-body integrals from our consideration. Then, the logical first step beyond the SA is using of an auxiliary basis (p') for the RI

$$|\kappa\rangle\langle\kappa| \approx |p'\rangle\langle p'| \tag{15}$$

This has been initially implemented at the MP2-R12 level [7–9], however, the ansatz for the $\mathcal{R}$ still remained $\mathcal{R} = \mathcal{R}_2$. Also, GBC and EBC, may not be required in that case. Nevertheless, results with smaller computational bases are not convincing.

Here, one has to note that as soon as GBC is not valid, the full ansatz of Eq. (11) should be used. In fact, inclusion of the "r_{12}-single-excitations is not a difficult task since one can easily transform the Hamiltonian as

$$H_{R12^1} = e^{-\mathcal{R}_1} H_N e^{\mathcal{R}_1} \tag{16}$$

and use the usual CCSD-R12 code with matrix elements of H_{R12^1} instead of H_N. Concerning the coding, much more demanding is to evaluate the necessary matrix elements involving the full $\mathcal{R}_2$ using the auxiliary basis. In the next section, we show some preliminary results calculated at the MP2-R12 level, as the initial approximation to CCSD-R12. Matrix elements including the $\mathcal{R}_2$ were calculated as given in Ref. [10]. Now the energy reads:

$$E^{(2)} = \langle\Phi|\hat{H}_N\hat{T}|\Phi\rangle + \langle\Phi|W_N\hat{\mathcal{R}}_2|\Phi\rangle + \langle\Phi|\hat{F}_N\mathcal{R}_1|\Phi\rangle = E_{conv}^{(2)} + E_{R12}^{(2)} \tag{17}$$

$$E_{R12^1}^{(2)} = f_k^\alpha \bar{r}_{\alpha j}^{ij} c_i^k \approx f_k^{p''} \bar{r}_{p''j}^{ij} c_i^k; \qquad p'' = p' - p \tag{18}$$

Pertinent c amplitudes are obtained by solving set of equations:

$$\langle\Phi|\tilde{\mathcal{R}}_k^i \hat{F}_N(1 + \mathcal{R}_1)|\Phi\rangle^{(1)} = 0, \tag{19}$$

superscript "(1)" meaning that only first order terms are to be considered. Using the auxiliary basis gives rise to:

$$-\bar{r}_{ij}^{p''j} f_{p''}^{q''} \bar{r}_{q''n}^{mn} c_m^k + \bar{r}_{ij}^{p''j} r_{p''n}^{mn} c_m^o f_o^k + \bar{r}_{ij}^{p''j} f_{p''}^k = 0 \tag{20}$$

III. MP2-R12 RESULTS

TABLE 1: MP2-R12 energies and their components for CO and H_2O molecules.

Basis	SCF+$E^{(2)}/E_h$	$E^{(2)}_{conv}/E_h$	$E^{(2)}_{R12}/E_h$	$E^{(2)}_{R12^1}/E_h$
	CO			
A-f	-113.308709	-0.489927	-0.028195	-0.000183
A-g	-113.310023	-0.501680	-0.017553	-0.000012
A-h	-113.310532	-0.505460	-0.014265	0.000000
cc-pVDZ	-113.193689	-0.290932	-0.153529	-0.010092
cc-pVTZ	-113.260317	-0.379942	-0.100101	-0.002177
cc-pVQZ	-113.291441	-0.437195	-0.065478	-0.000272
cc-pV5Z	-113.302523	-0.460168	-0.051822	-0.000070
	H_2O			
A-fp	-76.427744	-0.337640	-0.023089	-0.000309
A-fd	-76.428463	-0.340967	-0.020164	-0.000064
A-gd	-76.429234	-0.348649	-0.013199	-0.000026
A-gf	-76.429293	-0.349343	-0.012550	-0.000014
A-hd	-76.429445	-0.351512	-0.010519	-0.000004
A-hf	-76.429492	-0.351827	-0.010245	0.000000
cc-pVDZ	-76.327201	-0.203984	-0.096413	-0.004495
cc-pVTZ	-76.389706	-0.275096	-0.057452	-0.001122
cc-pVQZ	-76.414002	-0.313305	-0.035875	-0.000369
cc-pV5Z	-76.423124	-0.328787	-0.027261	-0.000060
cc-pV6Z	-76.426670	-0.337932	-0.021346	-0.000015

In Table 1 we show the convergence towards the basis set limit energy for two hierarchies of basis sets. First, results with R12 optimized $19s14p8d6f4g3h$ for non-hydrogen [11] and $9s6p4d3f$ for hydrogen [12] and their subsets are reported (A). The highest included angular momenta are indicated in the first column. The largest sets with all angular momenta included were always used as the auxiliary bases. Needless to say that in this case the basis set is close to the HF limit for atoms and contributions from the r_{12}-single-excitations are very small. In the second part, we have used the cc-pVXZ hierarchy for computational basis, while the auxiliary basis remained almost as before. Since in our formulation the auxiliary basis must contain the computational one, we have always used the unification of both from which the linear dependencies have been removed. With this hierarchy, the r_{12}-single-excitations contribution is by far not negligible.

Acknowledgment

This work was supported by the Slovak Grant Agency (Contract No. 2/6182/26).

---

[1] Noga, J.; Kutzelnigg, W. J Chem Phys, 1994, 101, 7738.
[2] Klopper, W.; Noga, J. ChemPhysChem, 2003, 4, 32.
[3] Boys, S. F.; Handy, N. C. Proc. Roy. Soc. A 1969, 310, 43.
[4] Čížek, J.; Adv Chem Phys, 1969, 14, 35.
[5] Kutzelnigg, W., Theor Chim Acta, 1985, 68, 445.
[6] Kutzelnigg, W.; Klopper, W. J Chem Phys, 1991, 94, 1985.
[7] Klopper, W.; Samson, C. C. M. J Chem Phys, 2002, 116, 6397.
[8] Ten-no, S.; Manby, F. R. J Chem Phys, 2003, 119, 5358.
[9] Valeev, E. F. Chem Phys Lett, 2004, 395, 190.
[10] Kedžuch, S.; Milko, M.; Noga, J. Int J Quantum Chem, 2005, 105, 929.
[11] Kedžuch, S.; Noga, J.; Valiron, P. Mol Phys, 2005, 103, 999.
[12] Noga, J.; Valiron, P. Collect Czech Chem Commun, 2003, 68, 340.

E. A. G. Armour, J. Franz and J. Tennyson (eds.)
Explicitly Correlated Wavefunctions

Accurately solving the electronic Schrödinger equation using explicitly correlated multi-reference methods

Robert J. Gdanitz†

North Carolina A&T State University, Physics Department, Rm. 101, Marteena Hall, Greensboro, NC 27411, U.S.A.

The (explicitly correlated) r_{12}-MR-ACPF (multi-reference averaged coupled-pair functional) method (and the novel more stable variant ACPF-2) are capable of closely reaching the full-CI and the basis set limits at the same time. These methods can be applied to entirely general chemical systems and they are available in the free AMICA (Atoms & Molecules In Chemical Accuracy) software which has recently been interfaced to ECCE (Extensible Computational Chemistry Environment). In a series of pioneering applications to difficult problems, i.e. the dissociation of N_2, the total energies of first-row atoms and positive and negative atomic ions, the molecular constants of He_2, the treatment of Be_2, the dispersion interaction in Ne_2, the excited valence states of CH_2, and the potential energy surface of the elementary reaction $F + H_2$ along the internal reaction coordinate, r_{12}-MR-ACPF gives accuracies that are either close to those of (sometimes highly specialized) very high-level methods or better. Possible future enhancements of the present method are briefly discussed.

I. INTRODUCTION

We wish to solve the electronic Schrödinger equation,

$$\hat{H}(\mathbf{r};\,\mathbf{R})\,\Psi(\mathbf{r};\,\mathbf{R}) = E(\mathbf{R})\,\Psi(\mathbf{r};\,\mathbf{R}) \tag{1}$$

---

†On leave from: *Institute for Physical and Theoretical Chemistry, Technical University Braunschweig, Hans-Sommer-Str. 10, D-38106 Braunschweig, Germany* and *Henry-Eyring-Center for Theoretical Chemistry, University of Utah, 315 S. 1400 E. Rm. 2020, Salt Lake City, UT 84112, U.S.A.*

for low-lying valence states. $\hat{H}$ is the spin-free, non-relativistic clamped-nuclei Hamiltonian and $\mathbf{r}$ and $\mathbf{R}$ are the (spatial) coordinates of the electrons and the nuclei, respectively. The wavefunction Ψ is constrained to be anti-symmetric with respect to the exchange of two electrons,

$$\Psi(\ldots, \mathbf{r}_i, \ldots, \mathbf{r}_j, \ldots) = -\Psi(\ldots, \mathbf{r}_j, \ldots, \mathbf{r}_i, \ldots) \tag{2}$$

The energy eigenvalue, E, only depends on the coordinates of the nuclei and, for a polyatomic system, gives rise to a potential energy surface (PES).

In molecular orbital theory, the wavefunction is expanded into Slater determinants, Φ_i,

$$\Psi \approx \tilde{\Psi} := \sum_k c_k \Phi_k \tag{3}$$

which are given as anti-symmetrized products of spin orbitals, φ_i,

$$\Phi_k(\mathbf{r}, \sigma) := \hat{\mathcal{A}}_N \prod_{i=1}^{N} \varphi_{i_k}(\mathbf{r}_i) \equiv \hat{\mathcal{A}}_N \, \varphi_{1_k}(\mathbf{r}_1) \cdots \varphi_{N_k}(\mathbf{r}_N) \tag{4}$$

where $\hat{\mathcal{A}}_N$ is the N electron antisymmetrizer, which enforces Eqn. (2). The (spin) orbitals are expanded into n nuclei-centered basis functions, χ_μ,

$$\varphi_i = \sigma_i \sum_{\mu=1}^{n} c_{\mu i} \chi_\mu \tag{5}$$

with the spin functions

$$\sigma_i \in \{\alpha, \beta\} \tag{6}$$

In praxi the basis functions, χ_μ, have to be taken from a finite (and therefore incomplete) n-dimensional subset of the one-electron Hilbert space. Additionally, the (full) configuration interaction (CI) expansion, Eqn. (3), consists of as many as $\binom{2n}{N}$ individual terms and, with a meaningful basis set, the former expansion needs to be truncated for chemical systems that have more than, say, four electrons. Upon truncation, the CI, Eqn. (3), c.f. Sect. II, and the basis set expansions, Eqn. (5), c.f. Sect. III, both converge rather slowly to the respective limit.

II. THE MULTI-REFERENCE AVERAGED COUPLED-PAIR FUNCTIONAL (MR-ACPF)

In so-called single-reference cases, where the wavefunction, Ψ, is dominated by a single term, Ψ_0, which usually is of Hartree-Fock type, the coupled-cluster

(CC) method, where the wavefunction is expanded into an exponential,

$$\tilde{\Psi} = \exp\left(\sum_{i,a} t_i^a \hat{a}_i^a + \sum_{\substack{a<b\\ i<j}} t_{ij}^{ab} \hat{a}_{ij}^{ab} + \dots\right) \Psi_0 \equiv \exp(\hat{T})\, \Psi_0 \tag{7}$$

with

$$\hat{T} := \sum_{i=1}^{N} \hat{T}_i \tag{8}$$

provide a rather fast converging truncation of the full-CI expansion, Eqn. (3). The $\hat{a}_{ij\dots}^{ab\dots}$ are excitation operators that substitute the spin-orbitals i, j, ... in Ψ_0 by virtual orbitals a, b, ... and the $t_{ij\dots}^{ab\dots}$ are the corresponding so-called cluster amplitudes. When the CC method is applicable, the computed energy converges that fast with respect to a truncation of the cluster expansion at a given level i [c.f. Eqn. (8)] that usually a truncation to the level of singles, doubles, and triples (CCSDT) or even singles and doubles (CCSD) is sufficiently accurate. In contrast to the truncation of the CI expansion, Eqn. (3), the coupled cluster expansion, Eqn. (7), is *size extensive* with respect to the truncation of $\hat{T}$ at any cluster level. We recall that size extensivity means that upon extension of a chemical system with non-interacting sub-systems, the computed energy grows linearly with the size of the system,

$$E(\mathrm{AB}\dots) = E(\mathrm{A}) + E(\mathrm{B}) + \dots \tag{9}$$

Unfortunately, CC theory turns out to be difficult to extend to the (more general) so-called multi-reference (MR) case where the wavefunction is dominated by a sum of a few Slater determinants,

$$\Psi \approx \Psi_0 = \sum_i c_i \Phi_i \tag{10}$$

the so-called references which all have non-negligible weights. MR ansätze are mandatory for the treatment of, e.g., covalent bonds that are elongated from their equilibrium value and finally broken, of excited states with the same spin and spatial symmetry as a lower lying state, and, in general, of all cases of near degeneracies. To the best knowledge of the author, no rigorous coupled cluster theory that allows for large-scale calculations of general chemical systems exists.

When the full-CI expansion, Eqn. (3), is truncated to the singles and doubles level, however, it is possible to obtain an approximately size-extensive theory by applying an *ad hoc* modification to the energy functional. To this end we write

$$E = E_0 + E_c, \quad E_0 := \langle \Psi_0 | \hat{H} | \Psi_0 \rangle \tag{11}$$

and intermediately normalize the wavefunction,

$$\tilde{\Psi} = \Psi_0 + \Psi_c, \quad \langle \Psi_0 | \Psi_c \rangle = 0, \quad \langle \Psi_0 | \Psi_0 \rangle = 1 \tag{12}$$

With the zeroth-order wavefunction optimized, e.g. on the multi-configurational self-consistent field (MCSCF) level, and then fixed, the correlation energy takes the form

$$E_c[\Psi_c] := \frac{\langle \Psi_0 + \Psi_c | \hat{H} - E_0 | \Psi_0 + \Psi_c \rangle}{1 + g \langle \Psi_c | \Psi_c \rangle} \tag{13}$$

with $g = 1$. Stationarization of the energy,

$$\delta_1 E_c[\Psi_c] \overset{!}{=} 0 \tag{14}$$

leads to the condition

$$\hat{Q}(\hat{H} - E_0 - gE_c) | \Psi_0 + \Psi_c \rangle \overset{!}{=} 0, \quad \hat{Q} := 1 - |\Psi_0\rangle\langle\Psi_0| \tag{15}$$

A better choice for g can be found from the condition that, when the chemical system of interest is appended by identical, non-interacting sub-systems, the shift gE_c has to stay constant, i.e. g has to be inversely proportional to the size of the system. Additionally, for a 2-electron system, g should be equal to one since then CI(SD) is exact (within the given basis set). Thus we are led to

$$g = \frac{2}{N} \tag{16}$$

Eqn. (13) and (16) then define the averaged coupled-pair functional (ACPF) [1].

In praxi it turns out that the ACPF tends to overestimate the electron correlation effect and it may even become instable. This is, to a large extend, due to an incorrect handling of the singles. In coupled-cluster theory, the correlation energy is given by

$$E_c^{\mathrm{CC}} = \langle \Psi_0 | (\hat{H} - E_0)(\hat{T}_1 + \tfrac{1}{2}\hat{T}_1^2 + \hat{T}_2) | \Psi_0 \rangle \tag{17}$$

In the ACPF, however, the term $\frac{1}{2}\hat{T}_1^2$ is (incorrectly) absorbed into the doubles, $\hat{T}_2$,

$$E_c^{\mathrm{ACPF}} = \langle \Psi_0 | (\hat{H} - E_0)(\hat{T}_1 + \hat{T}_2) | \Psi_0 \rangle \tag{18}$$

When the amplitudes of the singles are rather small—as is the case when Ψ_0 is a good zeroth-order wavefunction—the effect is small. When the singles are important because of deficiencies of Ψ_0, however, then ACPF may become instable. A simple remedy is to dampen the singles contribution by the choice

$$g' := \frac{4}{N}\left[1 - \frac{1}{2(N-1)}\right] \tag{19}$$

For $N \to \infty$, g' approaches $4/N = 2g$ and in the low-N limit, i.e. for $N = 2$, we have $g' = 1$, as in ACPF [and CI(SD)]. This modified version was recently proposed as ACPF-2 [2] and, in difficult cases as e.g. FeO [3], it proved to be considerably more stable than (the original) ACPF. When Ψ_0 is well-defined, however, ACPF and ACPF-2 give nearly identical results.

III. EXPLICITLY CORRELATED MULTI-REFERENCE METHODS

Since the wavefunction is symmetric with respect to the exchange of two electrons of opposite spins (singlet pair), the exact solution of the Schrödinger Eqn. (1) is *linear* in the interelectronic distance, r_{12},

$$\lim_{r_{12}\to 0}\left(\frac{\partial \Psi}{\partial r_{12}}\right)_{\text{av}} = \left.\tfrac{1}{2}\Psi\right|_{r_{12}=0} \tag{20}$$

("av" stands for the spherical average) [4], which gives rise to a cusp where two singlet electrons come close to each other. Since

$$r_{12} := |\mathbf{r}_1 - \mathbf{r}_2| = \sqrt{(x_1 - x_2)^2 + (y_1 - y_2)^2 + (z_1 - z_2)^2} \tag{21}$$

cannot (in a closed form) be written as a product $f(\mathrm{r}_1)g(\mathrm{r}_2)$, it is not surprizing that the CI expansion, Eqn. (3), which consists of anti-symmetrized products of one-electron functions, Eqn. (4), has difficulties to reproduce the cusp regions.

Eqn. (20) is obviously fulfilled when terms of the form

$$\tilde{\Phi}_{ij,k} := \hat{\mathcal{A}}_N\, r_{ij}\; \varphi_{1_k}(\mathbf{r}_1)\cdots\varphi_{i_k}(\mathbf{r}_i)\cdots\varphi_{j_k}(\mathbf{r}_j)\cdots\varphi_{N_k}(\mathbf{r}_N) \tag{22}$$

are added to the CI expansion. It is sufficient to construct these linear r_{12}-terms from the most important orbitals only. In general these are the internal orbitals that are required for a correct zeroth-order description of the wavefunction. In the spirit of the r_{12}-method of Kutzelnigg and Klopper [5,6] we remove those contributions from the r_{12}-terms that are already included in the CI expansion by constructing a projector $\hat{Q}'$,

$$\hat{Q}' := 1 - \sum_k |\Phi_k\rangle\langle\Phi_k| \tag{23}$$

where the sum runs over all determinants of the full-CI expansion. Application of $\hat{Q}'$ to the linear r_{12}-terms, Eqn. (22), leads to [7]

$$\hat{Q}' \, \tilde{\Phi}_{ij,r} \approx \tilde{\tilde{\Phi}}_{ij,r} := \left(\sum_{\kappa<\lambda} \bar{r}^{ij}_{\kappa\lambda} \, \hat{a}^{\kappa\lambda}_{ij} - \sum_{p<q} \bar{r}^{ij}_{pq} \hat{a}^{pq}_{ij} \right) \Phi_r \tag{24}$$

with p, q and κ, λ being indices of orbitals taken from the given and a (hypothetical) complete orthonormal basis set, respectively, and

$$\bar{r}^{ij}_{\kappa\lambda} := \int\!\!\int \left[\varphi_i^*(\mathbf{r}_1)\, \varphi_j^*(\mathbf{r}_2) - \varphi_i^*(\mathbf{r}_2)\, \varphi_j^*(\mathbf{r}_1)\right] r_{12} \; \varphi_\kappa(\mathbf{r}_1)\, \varphi_\lambda(\mathbf{r}_2)\, d\mathbf{r}_1 \, d\mathbf{r}_2 \tag{25}$$

The terms that are neglected in Eqn. (24) lead to singly substituted determinants and all matrix elements over these latter terms vanish within the so-called standard approximation. This approximation requires that the given basis set is "saturated" up to 3 $L_{\max}$, where $L_{\max}$ is the highest angular momentum number that occurs in the given system within the separated atoms limit. E.g. for a system consisting of first-row atoms, the basis set has to be saturated up to $L = 3$ since we have p-orbitals ($L = 1$) at most. For the treatment of molecules, it is important that the wavefunction is invariant with respect to a unitary transformation of (equivalent) internal orbitals and to this end we include off-diagonal terms as well [8],

$$\tilde{\tilde{\Phi}}{}^{kl}_{ij,r} := \left(\sum_{\kappa<\lambda} \bar{r}^{kl}_{\kappa\lambda} \hat{a}^{\kappa\lambda}_{ij} - \sum_{p<q} \bar{r}^{kl}_{pq} \hat{a}^{pq}_{ij} \right) \Phi_r \tag{26}$$

By applying the standard approximation, by analytical summations of infinite series, by algebraic reformulations, and by neglecting contributions that vanish fast enough with increasing L it is possible to derive expressions of the matrix elements of a r_{12}-MR-CI (and thus also of r_{12}-MR-ACPF) that can be evaluated using one- and two-electron integrals only. The derivations of these expressions are straightforward, although rather laborious, and for the explicit formulas we refer the reader to the literature [7,9].

IV. APPLICATIONS

The methods described in this paper can be applied to entirely general chemical systems and they are available in the AMICA [10] (Atoms & Molecules In Chemical Accuracy) program package, which is based on the COLUMBUS suite of MR-CI programs [11]. Very recently, AMICA has been implemented [12] into the Extensible Computational Chemistry Environment (ECCE) of PNNL [13] which allows for building a chemical system, preparing the input, submitting and monitoring the job, and finally viewing and archiving the results.

In a first series called "Accurately solving the Schrödinger equation of atoms and molecules using explicitly correlated (r_{12}-)MR-CI," the r_{12}-MR-ACPF method has been applied to the dissociation of N_2 (Part I) [14], to first-row atoms and positive ions (Part II) [15], to negative first-row atomic ions (Part III), [16], to He_2 (Parts IV and VI) [17], to Be_2 (Part V) [18], to HF (Part VII) [19] and finally to the valence excited states of methylene (CH_2, Part VIII) [20]. The r_{12}-MR-ACPF method has been used to obtain an accurate Ne_2 interaction potential [21], which has been applied to simmulations of the liquid and gaseaous phase of neon [22]. Very recently, the PES of the important elementary reaction F + $H_2 \rightarrow$ HF + H has been computed along the internal reaction coordinate [23].

V. POSSIBLE IMPROVEMENTS

The present methods, r_{12}-MR-ACPF and -ACPF-2, give rise to accuracies that are either close to those of (sometimes highly specialized) very high-level methods or better. Still, especially in the light of recent progress of approximate MR-CC theories and the r_{12}-method, there is space for improvement.

The MR-CEPA1 (coupled electron-pair approximation) method as well as an averaged version called MR-ACEPA, that have very recently been proposed by Ruttink and co-workers [24] give rise to energies that, in general, are close to MR-ACPF. In size extensivity tests that involved open shell molecules, the errors of the new methods, however, were found to be considerably smaller than the ones of MR-ACPF.

When the basis set is appended by individual functions at a time, energies that have been computed with the original formulation of the r_{12}-method, in general, do not converge smoothly to the limit (although L convergence is smooth with basis sets that are sufficiently close to saturation for each l-value). This behavior can be quite disturbing and it can be traced to errors of the resolution of the identity (RI). Indeed, in the original formulation of the r_{12}-method [6], as well as in the present formulation on the MR level [7] the basis set that is used to expand the wavefunction is also taken for the RI approximation. Recent investigations, however, indicate that it can be beneficial to use a different, so-called auxiliary basis set for the RI approximation. This also allows to expand the wavefunction in medium sized or, if desired, even in small basis sets. Recent developments are: i) the usage of an auxiliary basis set for the resolution of the identity in the one-electron space [25]; ii) density fitting of two-electron integrals [26]; and iii) the substitution of r_{12} by terms that more accurately reproduce the shape of the interelectronic cusp at intermediate distances, e.g. by exponentials or Gaussians (or linear combinations thereof) that depend on r_{12} and that are optionally multiplied by r_{12} [27]. The usage of auxiliary basis sets and correlation factors that more closely resemble the shape of the interelectronic cusp may also reduce or even remove disturbing instabilities due to inaccuracies of the matrix elements that now require special actions (c.f. [28,9]).

Work to improve these two shortcomings of the present method is underway.

---

[1] R. J. Gdanitz, and R. Ahlrichs. *Chem. Phys. Lett.* **143** (1988), 413–20.
[2] R. J. Gdanitz. *Intern. J. Quant. Chem.* **85** (2001), 281–300.
[3] W. Cardoen, and R. J. Gdanitz. *Chem. Phys. Lett.* **364** (2002), 39–43.
[4] T. Kato. *Commun. Pure Appl. Math.* **10** (1957), 151–77.
[5] W. Kutzelnigg. *Theoret. Chim. Acta* **68** (1985), 445–69.
[6] W. Kutzelnigg, and W. Klopper. *J. Chem. Phys.* **94** (1991), 1985–2001.
[7] R. J. Gdanitz. *Chem. Phys. Lett.* **210** (1993), 253–60.
R. J. Gdanitz, and R. Röhse. *Intern. J. Quant. Chem.* **55** (1995), 147–50. **59** (1996), E505.
[8] W. Klopper. *Chem. Phys. Lett.* **186** (1991), 583–5.
[9] R. J. Gdanitz. In *Recent Research Developments in Quantum Chemistry*, vol. 3. Transworld Research Network, Kerala, India, 2002, pp. 245–76.
[10] R. J. Gdanitz. Amica ("Atoms & Molecules In Chemical Accuracy") suite of r_{12}-MR-CI programs, unpublished, 1997. Amica is distributed freely through the Internet site `http://gdanitz.hec.utah.edu/amica`.
[11] H. Lischka, R. Shepard, F. B. Brown, and I. Shavitt. *Intern. J. Quant. Chem., Quant. Chem. Symp.* **15** (1981), 91–100.
R. Ahlrichs, P. Scharf, and C. Ehrhardt. *J. Chem. Phys.* **82** (1985), 890–8.
R. Shepard, I. Shavitt, R. M. Pitzer, D. C. Comeau, M. Pepper, H. Lischka, P. G. Szalay, R. Ahlrichs, F. B. Brown, and J.-G. Zhao. *Intern. J. Quant. Chem., Quant. Chem. Symp.* **22** (1988), 149–65.
R. Shepard, H. Lischka, P. G. Szalay, T. Kovar, and M. Ernzerhof. A general multireference configuration interaction gradient program. *J. Chem. Phys.* **96** (1992), 2085–98.
[12] R. J. Gdanitz, G. D. Black, C. S. Lansing, B. J. Palmer, and K. L. Schuchardt. *J. Comp. Chem.* **26** (2005), 214–25.
[13] G. Black, B. Didier, D. Feller, D. Gracio, M. Hackler, D. Jones, E. Jurrus, T. Keller, C. Lansing, S. Matsumoto, B. Palmer, K. Schuchardt, E. Stephan, H. Taylor, G. Thomas, E. Vorpagel, and T. Windus. "Extensible Computational Chemistry Environment (ECCE), A Problem Solving Environment for Computational Chemistry, Software Version 2.1 (2002)" as developed and distributed by Pacific Northwest National Laboratory,

P.O. Box 999, Richland, Washington 99352, USA, and funded by the U.S. Department of Energy (see `http://www.emsl.pnl.gov:2080/docs/ecce`).
[14] R. J. GDANITZ. *Chem. Phys. Lett.* **283** (1998), 253–61. **288** (1998), E590–2; **295** (1998) E540.
[15] R. J. GDANITZ. *J. Chem. Phys.* **109** (1998), 9795–801.
[16] R. J. GDANITZ. *J. Chem. Phys.* **110** (1999), 706–10.
[17] R. J. GDANITZ. *Mol. Phys.* **96** (1999), 1423–34.
R. J. GDANITZ. *Mol. Phys.* **99** (2001), 923–30.
[18] R. J. GDANITZ. *Chem. Phys. Lett.* **312** (1999), 578–84.
[19] W. CARDOEN, AND R. J. GDANITZ. *J. Chem. Phys.* **123** (2005), 024304.
[20] J. R. FLORES, AND R. J. GDANITZ. *J. Chem. Phys.* **123** (2005), 144316.
[21] R. J. GDANITZ. *Chem. Phys. Lett.* **348** (2001), 67–74.
[22] M. VENKATRAJ, M. G. MÜLLER, H. HUBER, AND R. J. GDANITZ. *Coll. Czech. Chem. Comm.* **68** (2003), 627–43.
M. VENKATRAJ, C. BRATSCHI, H. HUBER, AND R. J. GDANITZ. *Fluid Phase Equilibria* **218** (2004), 285–9.
[23] W. CARDOEN, R. J. GDANITZ, AND J. SIMONS. *Intern. J. Quant. Chem.* **106** (2006), 1516–27.
W. CARDOEN, R. J. GDANITZ, AND J. SIMONS. *J. Phys. Chem. A* **110** (2006), 564–71.
[24] P. J. A. RUTTINK, J. H. VAN LENTHE, AND P. TODOROV. *Mol. Phys.* **103** (2005), 2497–506.
P. J. A. RUTTINK. *Coll. Czech. Chem. Comm.* **70** (2005), 638–56.
[25] W. KLOPPER, AND C. C. M. SAMSON. *J. Chem. Phys.* **116** (2002), 6397–410.
W. KLOPPER. *J. Chem. Phys.* **120** (2004), 10890–5.
E. F. VALEEV, AND C. L. JANSSEN. *J. Chem. Phys.* **121** (2004), 1214–27.
H. FLIEGL, W. KLOPPER, AND C. HÄTTIG. *J. Chem. Phys.* **122** (2005), 084107.
S. KEDŽUCH, M. MILKO, AND J. NOGA. *Intern. J. Quant. Chem.* **105** (2005), 929–36.
[26] F. R. MANBY. *J. Chem. Phys.* **119** (2003), 4607–13.
S. TEN-NO, AND F. R. MANBY. *J. Chem. Phys.* **119** (2003), 5358–63.
H.-J. WERNER, AND F. R. MANBY. *J. Chem. Phys.* **124** (2006), 054114.
[27] A. J. MAY, AND F. R. MANBY. *J. Chem. Phys.* **121** (2004), 4479–85.
A. J. MAY, E. VALEEV, R. POLLY, AND F. R. MANBY. *Phys. Chem. Chem. Phys.* **7** (2005), 2710–3.
D. P. TEW, AND W. KLOPPER. *J. Chem. Phys.* **123** (2005), 074101.
F. R. MANBY, H.-J. WERNER, AND A. J. MAY. *J. Chem. Phys.* **124** (2006), 094103.
[28] W. CARDOEN, AND R. J. GDANITZ. *Comp. Meth. Sci. Tech.* **9** (2003), 31–43.

E. A. G. Armour, J. Franz and J. Tennyson (eds.)
Explicitly Correlated Wavefunctions
© 2006, CCP2, Daresbury

Recent advances in explicitly correlated electronic structure theory using short-range correlation factors

Seiichiro Ten-no

Graduate School of Information Science, Nagoya University, Chikusa-ku, Nagoya 464-8601, Japan
tenno@is.nagoya-u.ac.jp

I. INTRODUCTION

The slow convergence of orbital based expansions in the dynamic correlation aspect is the direct consequence of the inability to describe the cusp behavior [1–3] at coalescence with one-electronic basis functions. Various alternative methods have been proposed since the early success of the exploration of Hylleraas on the ground state of the helium atom [4]. The explicitly correlated methods turned out to be effective, yet they are handicapped by the introduction of many-electron integrals. The R12 methods [5,6] have transcended the limitation to tiny molecules by the systematic use of the resolution of identity (RI) to sidestep the explicit treatment of many-electron integrals. In this paper, recent advances in the field are described shortly mainly focusing on individual strategies by the present author.

II. THE HYLLERAAS ENERGY FUNCTIONAL AND ITS ROBUSTNESS

Henceforward, we use the following notations;

$i, j, \ldots$	occupied orbitals in the reference space.
$a, b, \ldots$	virtual orbitals in the given basis set.
$p, q, \ldots$	general orbitals in the given basis set.
$\alpha, \beta, \ldots$	virtual orbitals in the complete basis set.
$p', q', \ldots$	general orbitals in the auxiliary basis set assumed to be complete.

In the complete basis set limit, pair functions can be formally expressed as

$$u_{ij}^{(s)} = c_{ij}^{\alpha\beta}(s)\{\alpha\beta\}^{(s)} \quad , \tag{1}$$

where we used the Einstein summation convention, the curly bracket denotes spin-adapted pair functions, and s takes 0 and 1 for singlet and triplet, respectively. In the standard ab initio methods, $u_{ij}^{(s)}$ are approximated with a finite basis set,

$$\tilde{u}_{ij}^{(s)} = \tilde{c}_{ij}^{ab}(s)\{ab\}^{(s)} \ . \tag{2}$$

The second order energy

$$\tilde{E}^2 = (1+2s)\tilde{f}_{ij}^{(s)} \tag{3}$$

is obtained by minimizing the Hylleraas energy functional

$$\tilde{f}_{ij}^{(s)}(\tilde{u}_{ij}^{(s)}) = \langle \tilde{u}_{ij}^{(s)}|\hat{F}_1 + \hat{F}_2 - \epsilon_i - \epsilon_j|\tilde{u}_{ij}^{(s)}\rangle + 2\langle \tilde{u}_{ij}^{(s)}|r_{12}^{-1}|\{ij\}^{(s)}\rangle \geq \tilde{f}_{ij}^{(s)}(u_{ij}^{(s)}) \ , \tag{4}$$

which gives an upper bound of the correlation energy of the exact pair functions $u_{ij}^{(s)}$. The differentiations with respect to the coefficients give the relation for the true pair functions,

$$\langle \{\alpha\beta\}^{(s)}|(\hat{F}_1 + \hat{F}_2 - \epsilon_i - \epsilon_j)|u_{ij}^{(s)}\rangle + \langle \{\alpha\beta\}^{(s)}|r_{12}^{-1}|\{ij\}^{(s)}\rangle = 0 \ . \tag{5}$$

Thus the substitution of

$$\tilde{u}_{ij}^{(s)} = u_{ij}^{(s)} + \Delta_{ij}^{(s)} \tag{6}$$

shows that the energy functional is robust to involve only quadratic errors,

$$f_{ij}^{(s)}(\tilde{u}_{ij}^{(s)}) = f_{ij}^{(s)}(u_{ij}^{(s)}) + \langle \Delta_{ij}^{(s)}|(\hat{F}_1 + \hat{F}_2 - \epsilon_i - \epsilon_j)|\Delta_{ij}^{(s)} .\rangle \tag{7}$$

This means that the Hylleraas energy functional is not sensitive to a fine optimization of pair functions.

III. PAIR FUNCTIONS IN EXPLICILTY CORRELATED ELECTRONIC STRUCTURE THEORY

The cusp conditions [1–3] indicates that the exact pair functions behave as

$$u_{ij}^{(s)} = Q_{12}[\frac{r_{12}}{2(1+s)} + O(r_{12}^2)]\{ij\}^{(s)} \ , \tag{8}$$

where Q_{12} is the strong orthogonality projector,

$$Q_{12} = (1 - P_1)(1 - P_2) \ , \tag{9}$$

and P_n denotes the projection operator on the occupied space,

$$P_n = |\phi_i(n)\rangle\langle\phi_i(n)| \ . \quad (10)$$

The slow convergence of pair functions is originating from the cusps at coalescence. It is also noted that the cusp conditions are dependent on the spin indices of pair functions.

The explicitly correlated methods have mitigated this situation incorporating a correlation factor f_{12} dependent on inter-electronic distance directly as

$$\tilde{u}_{ij}^{(s)} = \tilde{c}_{ij}^{\prime ab}(s)\{ab\}^{(s)} + Q_{12}\tilde{c}_{ij}^{kl}(s)f_{12}\{kl\}^{(s)} \ . \quad (11)$$

In earlier methods with Gaussian-type geminals (GTGs), only the diagonal elements $\tilde{c}_{ij}^{ij}$ are dealt with as variational parameters. The method however leads to a trivial result for a super-molecule separated largely with an undamped correlation factor like the linear r_{12} function along with canonical orbitals. Klopper introduced the off-diagonal amplitudes in the unitary invariant formulation to remedy the problem [7]. Nevertheless, a clear indication is that the long-range component in the correlation factor is unphysical.

Because of the robustness of the Hylleraas energy functional, a possible compromise is to abandon all numerical parameters concerning the correlation factor but to employ the cusp conditions directly with a short-range factor $f_{12}^{(\text{short})}$ [8,9] as

$$\tilde{u}_{ij}^{(s)} = \tilde{c}_{ij}^{\prime\prime ab}(s)\{ab\}^{(s)} + Q_{12}\frac{1}{1+s}f_{12}^{(\text{short})}\{ij\}^{(s)} \ , \quad (12)$$

where the correlation factor is parameterized with respect to the s-wave cusp condition for singlet pairs [1–3],

$$f_{12}^{(\text{short})} = \frac{1}{2}r_{12} + O(r_{12}^2) \ . \quad (13)$$

The expression is still dependent on the spin index s, which can be removed by introducing the permutation operator to the spatial coordinates,

$$p_{12}\{pq\}^{(s)} = (-)^{(s)}\{pq\}^{(s)} \ . \quad (14)$$

The desired form [10] is

$$\tilde{u}_{ij}^{(s)} = \tilde{c}_{ij}^{\prime\prime ab}(s)\{ab\}^{(s)} + Q_{12}\tilde{f}_{12}^{(\text{short})}\{ij\}^{(s)} \ , \quad (15)$$

and "the rational generator" $\tilde{f}_{12}^{(\text{short})}$ from the short-range factor is given by

$$\tilde{f}_{12}^{(\text{short})} = \frac{1}{4}(3 + p_{12})f_{12}^{(\text{short})} \ . \quad (16)$$

The rational generator automatically imposes different correlation behaviors for different spins. Thus it is useful when figuring out the spin states embedded in a reference wave function is difficult, for instance in open shell situations.

IV. MANY ELECTRON INTEGRALS AND SLATER-TYPE CORRELATION FACTOR

Explicitly correlated methods introduce many-electron integrals as

$$\langle ij|r_{12}^{-1}P_2 f_{12}|kl\rangle = \sum_m \langle ijm|r_{12}^{-1}f_{13}|kml\rangle \ . \tag{17}$$

It was shown that all Schrödinger integrals which can arise in the use of correlated Gaussian functions can be calculated explicitly. The applicability is however limited to small molecules because of the many-electron integrals. Kutzelnigg and Klopper [6] introduced the innovative use of RI which breaks up many-electron integrals into sums of two-electron integrals,

$$\langle ij|r_{12}^{-1}P_2 f_{12}|kl\rangle = \sum_{p'm} \langle ij|r_{12}^{-1}|p'm\rangle\langle p'm|f_{12}|kl\rangle \ . \tag{18}$$

Although the original RI assumed the orbital basis set to be complete, the recent methods use an auxiliary basis set (ABS) [11] and their hybrid [12]. Consequently, the R12-type methods require only two-electron integrals over the operators like f_{12}, $f_{12}r_{12}^{-1}$, $[t_1+t_2, f_{12}]$, f_{12}^2, and $\frac{1}{2}[[t_1+t_2, f_{12}], f_{12}]$. The hybrid method requires two-electron integrals involving at most one index belonging to ABS. Thus the cost of the integrals scales as N^3N' for the numbers of basis set and ABS, N and N', respectively. The density fitting technique further improves the efficiency [13] and accuracy [9] of RI.

An attractive alternative to RI is simply to replace the sum over ABS by a numerical integration [10],

$$\langle ij|r_{12}^{-1}P_2 f_{12}|kl\rangle = \sum_{mg} w(\mathbf{r}_g)\phi_i(\mathbf{r}_g)\phi_k(\mathbf{r}_g)\langle j|r_{1g}^{-1}|m\rangle\langle m|f_{1g}|l\rangle \ , \tag{19}$$

where $w(\mathbf{r}_g)$ is the weights at the quadrature point $\mathbf{r}_g$. This expansion involves at most three-index objects, and the scaling for the necessary integrals is reduced to N^2G for the number of grid points G. The same idea was proposed by Boys and Handy long time ago within the framework of the transcorrelated methods [14]. It is noted that the fuzzy cell of Becke [15] makes the present scheme practical for molecules.

In either case of quadrature and RI, not many f_{12} lead to useful forms of the required two-electron integrals. However, any target function can be examined in explicitly correlated calculations by a fit with frozen GTG as first implemented by Persson and Taylor [16],

$$f_{12}^{(\mathrm{GTG})} = c_G \exp\left(-\zeta_G r_{12}^2\right) \ . \tag{20}$$

The linear r_{12} function is another popular form of correlation factor employed in the R12 methods extensively [5,6],

$$f_{12}^{(\mathrm{R12})} = \frac{1}{2} r_{12} \ . \tag{21}$$

Recently, we proposed the use of the Slater-type (exponential) correlation factor [17] in the form,

$$f_{12}^{(\mathrm{STG})} = -\frac{r_c}{2} \exp\left(-\frac{r_{12}}{r_c}\right) \ , \tag{22}$$

where r_c is a scale length parameter. Based on GTG fits, other authors have shown that $f_{12}^{(\mathrm{STG})}$ outperforms other correlation factors [18,19]. The most important feature of $f_{12}^{(\mathrm{STG})}$ is that all of the necessary integrals can be calculated efficiently in closed form algebraic expressions involving the special function,

$$G_m(T,U) = \int_0^1 dt t^{2m} \exp\left[-Tt^2 + U(1-t^{-2})\right] \ (m = -1, 0, 1, \ldots) \ . \tag{23}$$

Although it is not obvious if there exist a universal scale length parameter r_c, the reported results are not very sensitive to r_c [17,19].

V. MULTI-REFERENCE THEORY

The computational methods explained so far can be readily extended to the multi-reference (MR) framework. For instance, the standard MR perturbation (PT) wave function based on the complete active space self-consistent field (CAS-SCF) reference state [20] is expanded into the configuration state functions in the first order interacting space. They are generated by multiplying the spin-free excitation operators $\hat{E}_{ij}\hat{E}_{kl}$ (internal), $\hat{E}_{ij}\hat{E}_{ak}$ (semi-internal), and $\hat{E}_{ai}\hat{E}_{bj}$ (external) to the reference function, where we used the redefined indices, $i, j, \ldots$ denoting the active and inactive orbitals. The contributions of the different subspaces have different convergence properties. It is expected that the internal and semi-internal contributions saturate quickly for a fixed number of active orbitals without high angular momentum. Hence the simple form of the MR first order wave function is expressed as a sum of the components of the orbital expansion $|\Psi_O^{(1)}\rangle$ and of the excitation by the short-range correlation factor,

$$|\Psi^{(1)}\rangle = |\Psi_O^{(1)}\rangle + \hat{\Omega}_X^f |\Phi_{\mathrm{ref}}\rangle \ , \tag{24}$$

where $|\Phi_{\mathrm{ref}}\rangle$ is the reference wave function of CAS or its subset, and $\hat{\Omega}_X^f$ is the spin-free excitation operator of the rational generator in the external subspace,

$$\hat{\Omega}_X^f = \frac{1}{2} \langle \alpha\beta | \, \tilde{f}_{12}^{(short)} \, | ij \rangle \, E_{\alpha i} E_{\beta j} \ . \tag{25}$$

The matrix elements over the rational generator are simply expressed by

$$\langle\alpha\beta|\tilde{f}_{12}^{(\text{short})}|ij\rangle = \frac{3}{4}\langle\alpha\beta|f_{12}^{(\text{short})}|ij\rangle - \frac{1}{4}\langle\alpha\beta|f_{12}^{(\text{short})}|ji\rangle\,. \tag{26}$$

It is noted that the explicitly correlated part is constructed in the internally contracted fashion. In the state-specific MRPT, the zeroth order Hamiltonian is in the form [21],

$$\hat{H}_0 = \hat{\mathrm{P}}\hat{\mathrm{F}}\hat{\mathrm{P}} + (\hat{\mathrm{Q}} + \hat{\mathrm{R}})\hat{\mathrm{F}}(\hat{\mathrm{Q}} + \hat{\mathrm{R}})\,, \tag{27}$$

where $\hat{\mathrm{F}}$ is usually a spin-averaged Fock operator, and $\hat{\mathrm{P}}$, $\hat{\mathrm{Q}}$, and $\hat{\mathrm{R}}$ are projectors onto the reference function, its orthogonal components in the given basis set, and their residual, respectively. This choice, however, spoils the faborable commutator form concerning $\hat{H}_0$, reflecting the incompleteness of the current status of MRPT [22]. Anyway, the resulting energy expression by the minimization of the Hylleraas energy functional is the conventional MRPT energy plus a complete basis correction involving a sum over the states in the first order interacting space. A further simplification is possible by assuming

$$\hat{H}_0 = \hat{\mathrm{P}}\hat{\mathrm{F}}\hat{\mathrm{P}} + \hat{\mathrm{Q}}\hat{\mathrm{F}}\hat{\mathrm{Q}} + \hat{\mathrm{R}}\hat{\mathrm{F}}\hat{\mathrm{R}}\,, \tag{28}$$

which is relevant to the extended Brillouin-type condition. The use of the rational generator is advantageous in MR methods both in the formal and practical aspects.

Acknowledgement

ST gratefully acknowledges the Japan Science and Technology Agency (JST), CREST for funding.

---

[1] T. Kato 1957 *Commun. Pure Appl. Math.* **10** 151
[2] R. T. Pack and W. Byers-Brown 1966 *J. Chem. Phys.* **45** , 556
[3] W. Kutzelnigg and J. D. Morgan III, 1992 *J. Chem. Phys.* **96** , 4484
[4] E. A. Hylleraas, 1929 *Z. Phys.* **54** , 347
[5] W. Kutzelnigg, 1985 *Theor. Chim. Acta.* **68** , 445
[6] W. Kutzelnigg, W. Klopper, 1991 *J. Chem. Phys.* **94** , 1985
[7] W. Klopper, 1991 *Chem. Phys. Lett.* **186** , 583
[8] S. Ten-no, *Lecture Notes in Computer Science*, Springer, **2660**, 152 (2003)
[9] S. Ten-no and F. R. Manby, 2003 *J. Chem. Phys.* **119** , 5358

[10] S. Ten-no, 2004 *J. Chem. Phys.* **121** , 117
[11] W. Klopper and C. C. M. Samson, 2002 *J. Chem. Phys.* **116** , 6397
[12] W. Klopper, 2004 *J. Chem. Phys.* **120** , 10890
[13] F. R. Manby, 2003 *J. Chem. Phys.* **119** , 4607
[14] S. F. Boys, 1960 *Proc. Roy. Soc. London Ser. A* **258** 402
[15] A. D. Becke, 1987 *J. Chem. Phys.* **88** , 2547
[16] B. J. Persson, P. R. Taylor, 1996 *J. Chem. Phys.* **105** , 5915
[17] S. Ten-no, 2004 *Chem. Phys. Lett.* **398** , 56
[18] A. J. May, E. Valeev, R. Polly, F. R. Manby, 2005 *Phys. Chem. Chem. Phys.* **7** , 2710
[19] D. P. Tew and W. Klopper, 2005 *J. Chem. Phys.* **123**, 74101
[20] B. O. Roos, P. Linse, P. E. M. Siegbahn, M. R. A. Bloomberg, 1981 *Chem. Phys.* **66** , 197
[21] K. G. Dyall, 1995 *J. Chem. Phys.* **102** , 4909
[22] S. Ten-no, unpublished.